T0190230

Green Energy and Technology

More information about this series at http://www.springer.com/series/8059

Parimita Mohanty · Tariq Muneer
Mohan Kolhe

Editors

Solar Photovoltaic System Applications

A Guidebook for Off-Grid Electrification

Editors
Parimita Mohanty
The Energy and Resources Institute
Delhi
India

Mohan Kolhe
University of Agder
Kristiansand S
Norway

Tariq Muneer
Edinburgh Napier University
Edinburgh
UK

ISSN 1865-3529
Green Energy and Technology
ISBN 978-3-319-36295-3
DOI 10.1007/978-3-319-14663-8

ISSN 1865-3537 (electronic)

ISBN 978-3-319-14663-8 (eBook)

Printed on acid-free paper

Springer International Publishing AG Switzerland is part of Springer Science+Business Media (www.springer.com)

Foreword

Of the many forms of renewable energies, solar photovoltaic (SPV) technology is one of the most mature and there is an increasing demand for SPV installations both in grid-connected as well as in off-grid, stand-alone modes. While conventionally straightforward designs were used to set up off-grid PV-based systems for a wide range of applications, with the advent of smart techniques it is now possible to adapt a smart design approach for off-grid stand-alone solar PV systems.

This book simplifies the off-grid solar PV system design and engineering processes providing all the practical steps and information required for the complete design. Further, the book provides a pragmatic approach for effective selection of different components of the solar PV system, such as solar modules, storage batteries and inverters. Sample, practical case studies along with the steps and the criteria to be followed for designing solar photovoltaic systems for various applications (such as solar home systems, solar charging stations for lighting applications, solar DC microgrids, AC solar mini-grids and Smart Micro/Mini-Grids.) make the book useful, particularly for energy designers, system integrators and practitioners.

An attempt has been made to explain the content in simple terms and any technical issue has been explained in a lucid language to ensure wider accessibility. Appropriate case examples are provided where necessary to support the explanation and further references have been suggested for those looking for more detailed information on a given topic.

This edited volume, which has been developed as part of the 'Off-grid Access System in South Asia' project, is an effort by the authors to bring those practical approaches in designing, implementing and assessing the performance of various solar PV systems for off-grid applications.

I believe this volume will prove to be a valuable addition to the literature on solar PV system design and will assist in promoting mini-grid, micro-grid and off-grid electrification efforts and act as an important contribution to the United Nations' Decade of Sustainable Energy for All.

Prof. S.P. Gon Chaudhuri
Ashden Awardee
Member, State Planning Board
Government of Tripura and Chairman
of Ashden India Collective

Preface

In the context of off-grid electrification, the solar photovoltaic (solar PV) technology has emerged as one of the most dominant and preferred options. This is primarily because the solar PV technology has the versatility and flexibility for designing systems for different regions, especially for remote off-grid applications. While conventionally straightforward designs were used to set up off-grid PV-based systems in many areas for wide range of applications, with the advent of smart techniques it is now possible to adapt a smart design approach for the off-grid stand-alone solar PV system. A range of off-grid system configurations are possible, from the more straightforward design to the relatively complex, depending upon power and energy requirements, electrical properties of the load as well as on-site specificity and available energy resources.

Although several theoretical approaches exist in designing as well as assessing the performance of off-grid solar PV systems, they do not truly reflect the actual practice being followed and results obtained from ground do not often match with the theoretical guidance. The installation and operational condition in the actual field condition is quite different and thus the design of the entire PV system needs to be done based on the field parameters and performance needs to be assessed based on practical conditions.

This edited volume is an effort to bring those practical approaches in designing, implementing and assessing the performance of various solar PV systems for off-grid applications. This book is a part of the dissemination activity of a Research Council's UK funded project on off-grid electrification, called OASYS South Asia, which is a collaborative research work of five partner organizations, namely De Montfort University (DMU), Manchester University, Edinburgh Napier University from UK and The Energy and Resources Institute (TERI) and TERI University from India. The consortium has since 2009 undertaken a significant amount of research on rural electrification, especially off-grid electrification, in South Asia.

The volume contains seven chapters. The main purpose of this guidebook is to provide the state-of-the-art practical knowledge about solar PV-based systems and to present a complete guide for planning, design and implementation of solar PV systems for off-grid applications. This book provides the steps and the criteria to be followed for designing solar photovoltaic systems for various applications (such as solar home systems, solar charging stations for lighting applications, solar DC micro grids, AC solar mini-grids, Smart Micro/Mini-Grids, etc.) keeping the practical conditions into consideration. Although the solar PV system itself is not new, the advent of smart digital technologies has brought many new dimensions to the entire design of the PV system and this book suggests how to bring smarter design approaches and techniques for improving the efficiency, reliability and flexibility of the entire system.

In addition, the book brings insight into the actual performance of different components of solar PV systems and recommends the approaches to be followed for selecting the appropriate components for solar PV installations (what works and what does not work) based on their own lessons learned and challenges faced in the field. It also deals with project feasibility and techno-economic aspects.

The guidebook is written keeping the general users in mind, without much specific technical or subject-specific expertise. Attempt has been made to explain the content in simple terms and any technical terms have been explained in a lucid language to ensure wider accessibility. Appropriate case examples are provided where necessary to support the explanation and further materials have been suggested for those looking for more detailed information on a given issue.

As the editors of the volume we would like to thank all the contributors to this volume for their continued support and hard work. We also thank Elsevier for allowing us to reuse the materials published in their journals. We thank TERI, New Delhi, India for allowing us to use their diagrams here. We are also grateful to our respective institutes, TERI, Edinburgh Napier University and University of Agder, for encouraging us to edit this volume. Despite the support from different quarters, errors, if any, are ours.

We are grateful to the funding agencies, Engineering and Physical Science Research Council and Department of International Development of United Kingdom, and their support to this initiative.

We also thank the publisher—Springer for agreeing to publish this volume despite the specialized nature of the work that still faces limited academic attention.

In particular, we would like to extend our sincere thanks to Prof. Subhes C. Bhattacharyya, the OASYS overall research coordinator and Series Editor of the project-related monographs for his terrific support throughout.

Last but not least, we would like to thank our respective families for their support in completing this work.

New Delhi, India	Parimita Mohanty
Edinburgh, UK	Tariq Muneer
Grimstad, Norway	Mohan Kolhe
July 2015	

Contents

Introduction 1
Subhes Bhattacharyya

Solar Radiation Fundamentals and PV System Components 7
Parimita Mohanty, Tariq Muneer, Eulalia Jadraque Gago
and Yash Kotak

PV System Design for Off-Grid Applications 49
Parimita Mohanty, K. Rahul Sharma, Mukesh Gujar, Mohan Kolhe
and Aimie Nazmin Azmi

PV Component Selection for Off-Grid Applications 85
Parimita Mohanty and Mukesh Gujar

Performance of Solar PV Systems 107
Tariq Muneer and Yash Kotak

Economics and Management of Off-Grid Solar PV System 137
K. Rahul Sharma, Debajit Palit and P.R. Krithika

**Hybrid Energy System for Rural Electrification in Sri Lanka:
Design Study** 165
Iromi Ranaweera, Mohan Lal Kolhe and Bernard Gunawardana

Introduction

Subhes Bhattacharyya

Abstract This chapter introduces the electricity access challenge and presents the decentralized approach to electrification of rural areas in developing countries. It then provides an overview of the content of the book.

1 Background

According to the World Energy Outlook 2014, nearly 1.3 billion people did not have access to electricity in 2012 and nearly 2.7 billion people used traditional biomass for cooking and other heating energy needs. Electricity did not reach to more than 620 million people in Sub-Saharan Africa and another 620 million suffered the same problem in developing Asia, with India alone accounting for more than 300 million without access to electricity (WEO 2014). More importantly, analysis reported in the World Energy Outlook 2013 indicates that although 1.7 billion are likely to be electrified until 2030, the growing population of developing countries could still leave another one billion without access to electricity in 2030 (WEO 2013). This critical situation has arisen despite growing appreciation of the role of energy in promoting sustainable development at the international development policy circle since the Johannesburg Summit in 2002.

Recognizing the challenge, the UN Secretary General, Mr. Ban Ki Moon, launched a global initiative, Sustainable Energy for All, aiming to provide universal access to energy by the year 2030. Year 2012 was declared as the "International Year of Sustainable Energy for All" and the UN General Assembly has unanimously decided to declare the decade 2014–2024 as the decade of Sustainable Energy for All. The energy access issue has climbed up the political agenda and more than 80 countries have joined the initiative.

S. Bhattacharyya (✉)
Institute of Energy and Sustainable Development, De Montfort University, Leicester, UK
e-mail: subhesb@dmu.ac.uk

© Springer International Publishing Switzerland 2016
P. Mohanty et al. (eds.), *Solar Photovoltaic System Applications*,
Green Energy and Technology, DOI 10.1007/978-3-319-14663-8_1

It is generally recognized that the top-down approach of grid extension remains the preferred mode of electrification but extension of the central electricity grid to geographically remote and sparsely populated rural areas can either be financially unviable or practically infeasible. This arises due to remoteness of the settlements from the existing grids, low-electricity demand with prominent evening peaks, low-population density and difficult terrain. Grid extension is capital intensive for remote and difficult terrains—the cost can vary between $6700 and $19,000 per kilometer (ARE 2011). IEA (2011) estimates that to achieve the universal electrification objective by 2030, in terms of technology choice, grid extension will cater for 30 % of the cases, whereas the rest 70 % will come from mini-grids or off-grid systems in the proportion of 65:35. Szabo et al. (2011) using a spatial least-cost analysis framework identified that in many parts of Africa cost of decentralized off-grid options can be cheaper than grid extension and that if the affordability of consumers can be increased or cost of supply is reduced, off-grid options can surely play an important role. In a similar study, Bazilian et al. (2012) also suggest that to provide universal basic electricity access, most rural areas in Africa will need off-grid supply systems, either based on diesel generators or solar PV systems.

Moreover, often the desired outcomes do not always materialize. For example, a study found that non-poor households benefitted most from rural electrification and that the productive use is rarely developed (World Bank 2008). Moreover, heavy reliance on subsidies makes the electrification process unsustainable. Electrification did not necessarily generate income generating opportunities and failed to spur economic development due to poor integration with rural development agenda.

As an alternative to grid extension, decentralized options have emerged in areas not served by the grid and/or where the quality of service is poor. These areas are inhabited by small communities who generally have low-disposable income, with irregular cash flows. Household-level solutions (such as solar lanterns and solar home systems) and collective applications for community needs (served through local grids or energy centres) have been attempted. Being a bottom-up approach, it captures the aspirations of the local population through a participatory process, although users bear additional risks and responsibilities. Studies show that decentralized systems respond quickly to market needs, mobilize suppliers, and consumers at the grassroots level and the private entrepreneur level, and are less influenced by bureaucratic inefficiencies, funding limitations and obsolete technologies. Decentralized systems reach previously neglected consumer groups, encourage local entrepreneurship, improve education and social life in underprivileged areas, and reduce pressure on big cities through reduced migration for economic purposes.

In the context of off-grid electrification, the solar photovoltaic (solar PV) technology has emerged as the dominant and preferred option. The global solar PV market has grown very rapidly in recent times, covering both grid-connected and off-grid applications. The dramatic price reductions of PV modules have contributed to this demand growth. PV system prices in the US market have declined by 6–7 % every year between 1998 and 2013 but the price fell by 12–15 % during 2012–2013 (Feldman et al. 2014). In addition, the modular size of the systems, low maintenance due to absence of any moving parts and reasonably high reliability have helped the technology to emerge as the leader.

1.1 Solar PV-Based Off-grid Electrification

Solar PV-based electrification has followed two main pathways:

a. individual systems catering to the needs of a single household and generally, the individual devices and systems offer the first step up the ladder. This category includes solar lanterns and solar home systems which can perform to a limited number of tasks but their small and modular size makes them affordable to the bottom of the pyramid consumers.
b. Collective systems—these are centralized local systems that provide electricity to a group of users, including households and commercial consumers. This category includes mini and micro grids either running in alternating current or direct current mode. These "mini-utilities" can support a larger set of activities, including energy used for productive activities. They are receiving greater attention in recent times, to complement grid-based electrification, as a possible electrification option for increasing connection rates in rural areas of developing countries.

Some additional details about these are given in Box 1.1.

Box 1.1: Basic details of some devices or systems used in PV-based electrification

Solar lantern: A solar lantern is a portable lighting device that houses either a CFL lamp or a LED-based luminaire in plastic or metal enclosure and contains a rechargeable battery and supporting electronic components. The battery is charged either using a separate PV module or is integrated with the lamp.

Solar home system: A typical SHS is composed of one or more PV modules, a battery bank and a charge controller. Solar PV modules charge the batteries during the day time and the stored energy supplies DC electricity to run appliances such as CFL/LED lamps, DC fan, TV, radio and even a small refrigerator. The energy inflow and outflow into and from the battery bank is controlled by the charge controller.

Solar charging station: These stations offer a battery or solar lantern charging service for a fee and use solar PV modules as the source of energy. The station is designed to facilitate plugging-in of lanterns or batteries and to ensure that all the lanterns and or batteries are adequately charged.

Solar DC micro grids: As the name suggests, they generate DC electricity using one or more solar panels and distribute power over a short distance using a local network. The supply normally takes place at 24 V and electricity is mainly used for lighting and mobile phone charging.

Solar mini-grids: These are local grids where electricity is generated centrally using an array of solar PV modules and distributed to households and small businesses using a low voltage distribution network. The power is generally supplied at 3-phase or single-phase AC supply at 220 V, 50 Hz. The electricity is stored in a battery bank and a power conditioning unit consisting of charge controllers, inverters, AC/DC distribution boards and necessary cabling is required to ensure reliable supply.

Source: Based on (Palit 2013)

Although both the pathways have seen growth in recent times, the individual devices appear to have a larger market share. Solar lanterns are becoming popular due to portability, ease of use, inherent safety and cleanliness, and low cost (IFC 2012). Some modern lanterns are also bundling mobile charging facility and a radio, offering greater value to users. However, the solar home systems (SHS) dominate the solar PV-based off-grid electrification. They can cater to a much wider set of services than solar lanterns as more power capacity is available from these systems. Households can operate a combination of lights, fans, televisions, radios and even other smaller appliances. The initial SHS were larger in size —50–150 Wp but only richer households could afford them and hence the market size was limited. Gradually, smaller the system size has appeared in the market to ensure penetration of these systems to poorer households. Now systems of 10 or 20 Wp are commonly available in many countries. The emergence of LED lights has also supported the miniaturization process, as the power demand has dramatically reduced. In addition, efficient DC appliances are also appearing for rural markets, making a wider range of services available to household users. Taking advantage of these developments, portable solar kits are emerging that do not need installation or much maintenance and are cheaper than the fixed SHS, making them affordable as well as acceptable as a product (IFC 2012).

However, the main limitation of the individual systems comes from the restricted scope of application, the maintenance burden on the users, lack of income generating opportunities and relatively high cost of installations. Generally, they do not allow electricity use for productive purposes and the user bears the responsibility of maintaining the system and replacing the components on expiry of life. Moreover, the individual systems are generally costly and even with innovative financing schemes these systems have hardly reached the lowest income class of the population.

A village-scale system on the other hand is likely to be cheaper than the sum of all individual systems in a village. The economies of scale and scope can work here to provide a comparable service, if not better. These mini-utilities come in various sizes and can use different energy technologies but solar PV-based micro and mini-grids have emerged as the market leader in recent times. If the users live in a close cluster and require a basic level of supply, a solar DC micro-grid can be a cost effective option. When the consumers are somewhat spread and the demand for productive or commercial load exists, a solar AC mini-grid can be used. In this electrification option, the service provider bears the responsibility of maintaining the system and the user avails the service for a fee.

Although several theoretical approaches exist in designing as well as assessing the performance of off-grid solar PV systems, they do not truly reflect the actual practice being followed and results obtained from ground do not often match with the theoretical guidance. The installation and operational condition in the actual field condition is quite different and thus the design of the entire PV system needs to be done based on the field parameters and performance needs to be assessed based on practical conditions. This area has not received adequate attention so far.

2 Purpose and Coverage of This Book

The main purpose of this guidebook is to provide the state-of-the-art practical knowledge about solar PV-based electrification systems and to present a complete guide for planning, design and implementation of solar PV systems for off-grid applications. This book provides the steps and the criteria to be followed for designing solar photovoltaic systems for various applications (such as solar home systems, solar charging stations for lighting applications, solar DC micro grids, AC solar mini-grids, and smart micro/mini-grids, etc.) keeping the practical conditions into consideration. Again, although the solar PV system itself is not new, the advent of smart digital technologies has brought many new dimensions to the entire design of the PV system and this book suggests how to bring smarter design approaches and techniques for improving the efficiency, reliability and flexibility of the entire system.

In addition, the book brings the insight into the actual performance of different components of solar PV systems and recommends on the approaches to be followed for selecting the appropriate components for solar PV installations (what works and what does not work) based on their own lesson learned and challenges faced at the field. It also deals with project feasibility and techno-economic aspects.

The guidebook is written keeping the general users in mind, without any specific technical or subject-specific expertise. An attempt has been made to explain the content in simple terms and any technical topic has been explained in a lucid language to ensure wider accessibility. Appropriate case examples are provided where necessary to support the explanation and further materials have been suggested for those looking for more detailed information on a given issue.

The book is divided into three sections: section A contains this introduction and an introduction to the trends in PV technology; section B deals with PV system design and system control while the last section covers the monitoring issues, project economics and management. A brief introduction to these chapters is given below.

In Chap. 2, Mohanty, Muneer, Gago and Kotak present an overview of PV technologies for off-grid electrification. This chapter also focuses on the trends of solar PV technologies, battery technologies, power converters and solar radiation measurements which plays the most critical role in a PV system

In Chap. 3, Kolhe, Mohanty, Sharma and Gujar provide a step-by-step guide on PV system design for off-grid electrification. This chapter covers the wide range of systems including solar charging stations, DC micro-grids, solar multi-utility centres and mini-grids. Inclusion of smart features and the considerations for future grid integration of the off-grid systems are also discussed in this chapter.

In Chap. 4, Mohanty and Gujar discuss PV component selection for off-grid electrification applications. The selection of PV modules, batteries, power converters and protection devices is presented. The effect of incorrect component selection is also highlighted.

In Chap. 5, Muneer and Kotak share their experience of PV system performance and present the results of performance testing of PV technologies, devices (such as solar lighting systems) and system such as batteries.

Sharma, Palit and Krithika present the techno-economics of solar PV-based off-grid electrification in Chap. 6. It demonstrates a methodology for calculation of the cost of generation of electricity from an off-grid solar PV power plant and showcases how business models are being developed in off-grid solar PV projects. It introduces the basic financial analysis concepts, demonstrates the estimation of levelized costs and shares experience from different case examples.

In Chap. 7, Ranaweera, Kolhe and Gunawardana present a design case study.

References

ARE. (2011). *Hybrid mini-grids for rural electrification: Lessons learned*. Brussels: Association for Rural Electrification.

Bazilian, M., Nussbaumer, P., Rogner, H. H., Brew-Hammond, A., Foster, V., Pachauri, S., Williams, E., Howells, M., Niyongabo, P., Musaba, L., Gallachoir, B. O., Radka, M., & Kammen, D. M. (2012). Energy access scenarios to 2030 for the power sector in Sub-Sahara n Africa. *Utilities Policy, 20*(1): 1–16.

Feldman, D., Barbose, G., Margolis, R., James, T., Weaver, S., Garghouth, N., et al. (2014). *Photovoltaic system pricing trends: Historical, recent and near-term projections*. Golden, CO: National Renewable Energy Laboratory.

IFC. (2012). *From gap to opportunity: Business models for scaling up energy access*. Washington, DC: International Finance Corporation.

Palit, D. (2013). Solar energy programs for rural electrification: Experiences and lessons from South Asia. *Energy for Sustainable Development, 17*(3), 270–279.

Szabo, S., Bodis, K., Huld, T., & Moner-Girona, M. (2011). Energy solutions in rural Africa: mapping electrification costs of distributed solar and diesel generation versus grid extension, *Environmental Research Letters, 6*, pp. 1–9, IOP Publishing, U.K.

WEO. (2013). *World Energy Outlook*. Paris: International Energy Agency.

WEO. (2014). *World Energy Outlook 2014*. Paris: International Energy Agency.

World Bank. (2008). *The welfare impacts of rural electrification: a reassessment of the costs and benefits, an IEG impact evaluation*. Washington, DC: The World Bank.

Solar Radiation Fundamentals and PV System Components

Parimita Mohanty, Tariq Muneer, Eulalia Jadraque Gago
and Yash Kotak

Abstract Solar PV technology has emerged as one of the most matured and fast evolving renewable energy technologies and it is expected that it will play a major role in the future global electricity generation mix. Keeping the rapid development of the PV technology into consideration, this chapter systematically documents the evolution of solar PV material as well as the PV applications and PV markets. It also provides insight into the trend in batteries and inverters used for solar PV applications. Furthermore, a comparative analysis of different PV technologies and its development is summarized. The rest of the chapter aims at providing a comprehensive analysis of solar radiation measurement and modelling techniques to assess the availability of solar radiation at different locations. The chapter presents comprehensive information for solar energy engineers, architects and other practitioners.

P. Mohanty (✉)
The Climate Group, Nehru Place, New Delhi, India
e-mail: pmohanty@theclimategroup.org

T. Muneer
Edinburgh Napier University, 10 Colinton Road, Edinburgh EH 10 5DT, UK
e-mail: T.Muneer@Napier.ac.uk

E.J. Gago
University of Granada, School of Civil Engineering, Avenida Severo Ochoa S/N,
18071 Granada, Spain
e-mail: ejadraque@ugr.es

Y. Kotak
Heriot Watt University, Riccarton, Edinburgh EH 14 4AS, UK
e-mail: yk78@hw.ac.uk

© Springer International Publishing Switzerland 2016
P. Mohanty et al. (eds.), *Solar Photovoltaic System Applications*,
Green Energy and Technology, DOI 10.1007/978-3-319-14663-8_2

1 Trends in Solar PV Technologies

1.1 Introduction

The global photovoltaic (PV) market in 2013 witnessed a massive growth with 38.4 GW (up from 30 GW in 2012) of new capacity around the globe and 11 GW installed in Europe alone. The most important fact from 2013 was the rapid development of PV in Asia both in terms of PV deployment and PV manufacturing (REN 21, 2014).

The global photovoltaic (PV) market in 2004 was only 3.7 GW which significantly reached to 139 GW in 2013 and the major developing countries are Germany, China, Italy, Japan, United States, Spain, France, United Kingdom, Australia and Belgium (REN 21 2014).

Almost 11 GW of PV capacity was connected to the grid in Europe in 2013, compared to 17.7 GW in 2012 and more than 22.4 GW in 2011. For the first time since 2003 Europe lost its leadership to Asia in terms of new installations. As shown in Fig. 1, China was the top market in 2013 with 12.9 GW of which 500 MW represented off-grid systems. China was followed by Japan with 6.9 GW and the USA with 4.8 GW (REN 21 2014).

Germany was the top European market with 3.3 GW. Several other European markets exceeded the one GW mark: the UK (1.5 GW) and Italy (1.5 GW) (REN 21 2014).

Several European markets that performed well in the past went down in 2013, a consequence of political decisions to reduce PV incentives, Belgian installations went from 600 MW in 2012 to 215 MW in 2013, French went from 1,115 MW to 613 MW, and Danish went down from 300 MW to around 200 MW. Aside from

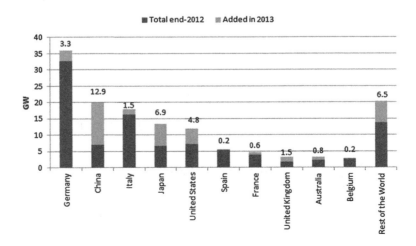

Fig. 1 Solar PV capacity and additions, top countries, 2013. *Source* REN 21 (2014)

the significant decline in Germany and Italy, the size of the remaining European PV market was stable, with around 6 GW per year in the last three years. Outside Europe, several markets continued to grow at a reasonable pace: India with 1,115 MW, Korea with 442 MW, Thailand with 317 MW and Canada with 444 MW in 2013 (EPIA 2014).

A point worth mentioning is that since the earthquake- and tsunami-related damage to the Fukushima plant in 2011, the Japanese energy policy is shifting away from nuclear power generation and this may indeed help deployment of PV on a much faster scale.

PV technology has the potential to contribute to at least 11 % of world's electricity supply by the year 2050 (IEA 2010). The PV module itself accounts for around half of total PV system cost. The costs of polysilicon and wafer production could decline dramatically by 2015 driven by the increasing scale of production and ongoing manufacturing innovations (see Fig. 2).

Building-integrated photovoltaic applications, commonly known as BIPV will have an increasingly important role within any given nation's energy matrix. In this respect the introduction of Feed-in Tariff (FIT) by more than 40 countries has given a major boost to the BIPV sector. Policies such as FIT will in the long term prove to be instrumental in bringing down the PV electricity generation costs to grid parity. A policy such as FIT promotes construction of high-quality systems and encourages PV financing and guaranteeing income over the life of the system. It also helps bringing common masses within the energy generators' group.

In March 2014, EPIA completed an extensive data collection exercise from a highly representative sample of the PV industry. EPIA derived three scenarios for the future development of PV markets (Masson et al. 2014).

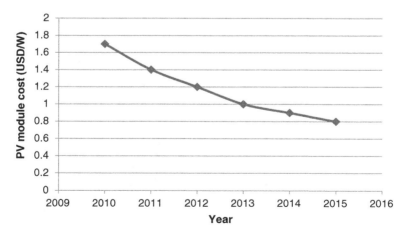

Fig. 2 The crystalline silicon PV module price projections for European, North American and Japanese manufacturers. *Data source* IRENA (2012)

- The High Scenario assumes the continuation, adjustment or introduction of adequate support mechanisms, accompanied by a strong political will to consider PV as a major power source in the coming years.
- The Low Scenario assumes rather pessimistic market behaviour with no major reinforcement or adequate replacement of existing support mechanisms, or a strong decrease/limitation of existing schemes or no adequate policies.
- The Medium Scenario weights the two previous scenarios according to the probability of achieving them.

As shown in Fig. 3, in the Low Scenario, the global market could remain between 35 and 39 GW annually in the coming five years. The combination of declining European markets and the difficulty of establishing durable new markets in emerging countries could cause this market stagnation.

MEA: Middle East and Africa. APAC: Asia Pacific.

In the High Scenario, the European market would first grow around 13 GW in 2014 before increasing slowly again to around 17 GW by year 2019. The global market could top more than 69 GW in 2018 (Fig. 4).

1.2 PV Technologies

Solar cell technologies in general are classified as wafer-based crystalline silicon solar cell technology, thin-film solar cell technology and other new emerging technologies as shown in Fig. 5.

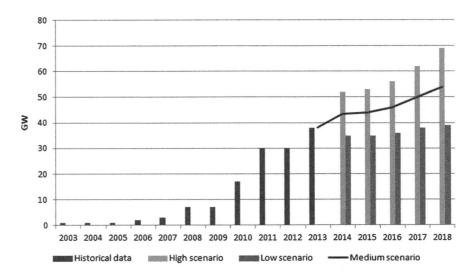

Fig. 3 Global annual market scenarios until 2018

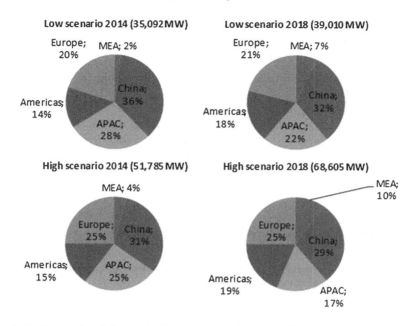

Fig. 4 Evolution of global annual PV market scenarios per region until 2018. *Source* Global market outlook for photovoltaics (2014–2018), EPIA

Fig. 5 Classification of solar cell technologies

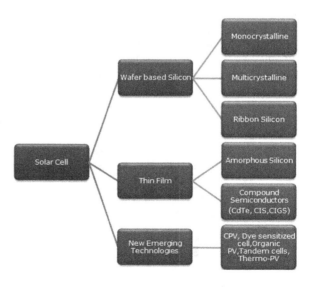

Out of the different solar PV technologies, there are two basic commercial PV module technologies available on the market today that are used by the Solar PV sector:

1. Wafer-based Solar cells made from crystalline silicon either as single or poly-crystalline wafers.
2. Thin-film products typically incorporate very thin layers of photovoltaic active material placed on a glass superstrate or a metal substrate using vacuum deposition manufacturing techniques similar to those employed in the coating of architectural glass. Like-with-like, commercial thin-film materials deliver roughly half the output of a thick crystalline PV array.

1.3   Market Share of Different PV Technologies

Thin-film production market share in the global solar PV market grew from a mere 5 % in 2005 to 33 % in 2014, thin-film solar PV, as shown in Fig. 6, is set to increase its share to 38 % by 2020 (Reporlinker 2011).

1.4   Trend in PV Technology Research

The public budgets for PV research and development in 2012 in the International Energy Agency Photovoltaic Power System Programme (IEA PVPS) countries are outlined in Table 1. The most significant reporting countries in terms of R&D funding are the USA, Germany, Korea, Japan, Australia, France and China. Governments are clearly identifying the benefits of this technology's further

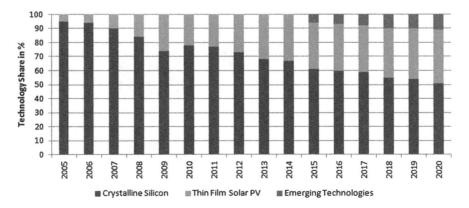

Fig. 6 Market share of the different PV technologies

Table 1 Public budgets for research and development in 2012 in the IEA PVPS countries

Countries	R&D in Mn USD	Increase from 2011
Austria	11.7 (2011)	N/A
Australia	26.9	−14 %
Denmark	4.32	−8 %
Canada	12	21 %
China	79	N/A
France	128 (3−5 years)	15 %
Germany	66	N/A
Italy	7.4	−8 %
Japan	130	28 %
Korea	118	26 %
Netherlands	35	40 %
Norway	14	N/A
Sweden	11.3	20 %
USA	262	Stable

Source IEA (2013)

development, better integration with existing energy systems and the benefits of innovation (IEA 2013).

With the aim of achieving further significant cost reductions and efficiency improvements, R&D is predicted to continuously progress in improving existing technologies and developing new technologies. In February 2011, Department of Energy (DOE) launched the SunShot Initiative, a programme focused on driving innovation to make solar energy systems cost-competitive with other forms of unsubsidized energy. In Germany, R&D is being conducted under the new 6th Programme on Energy Research 'Research for an Environmental Friendly, Reliable and Economically Feasible Energy Supply' which came into force in August 2011. Activities have three focal points: Organic solar cells, thin-film solar cells (with emphasis on topics such as material sciences including nanotechnology, new experimental or analytical methods), and the cluster called 'Solar valley Mitteldeutschland' in which most of the German PV industry participates. In Belgium, almost all PV technologies are studied: Organic, back-contact crystalline silicon, printed CIGS and CPV. Norway focuses mainly on the silicon value chain from feedstock to cells, but also fundamental material research and production processes (IEA 2013).

It is expected that a broad variety of technologies will continue to characterise the PV technology portfolio, depending on the specific requirements and economics of the various applications. Organic thin-film PV cells, using dye or organic semiconductors, have created interest and research, development and demonstration activities are underway. Figure 7 gives an overview of the different PV technologies and concepts under development.

As shown in Table 1, the main challenge for c-Si modules is to improve the efficiency and effectiveness of resource consumption through materials reduction,

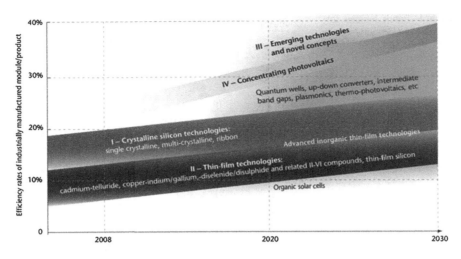

Fig. 7 Photovoltaic technology status and prospects. *Source* IEA Technology Roadmap, 2010

improved cell concepts and automation of manufacturing. Current commercial single c-Si modules have a higher conversion efficiency of around 14–20 %. Their efficiency is expected to increase up to 23 % by 2020 and up to 25 % in the longer term. Multi-crystalline silicon modules have a more disordered atomic structure leading to lower efficiencies, but they are less expensive. Their efficiency is expected to increase up to 21 % in the long term.

As shown in Table 2, the most promising R&D areas include improved device structures and substrates, large area deposition techniques, interconnection, roll-to-roll manufacturing and packaging. Similarly, the technology goals and key R&D issues for thin-film technologies are shown in Table 3.

Table 2 Technology goals and key R&D issues for crystalline silicon technologies

Crystalline silicon technologies	2010–2015	2015–2020	2020–2030/2050
Efficiency target redundant (commercial module)			
Single crystalline	21 %	23 %	25 %
Multi-crystalline	17 %	19 %	21 %
Industry manufacturing	Si consumption <5 g/W	Si consumption <3 g/W	Si consumption <2 g/W
R&D	New silicon materials and processing	Improved device structure; Cost optimization	New device structure with novel concepts

Source IEA Technology Roadmap (2010)

Table 3 Technology goals and key R&D issues for thin-film technologies

Thin-film technologies	2010–2015	2015–2020	2020–2030/2050
Efficiency target in % (commercial module)			
Thin-film silicon	10 %	12 %	15 %
CIGS	14 %	15 %	18 %
CdTe	12 %	14 %	15 %
Industry manufacturing	Roll-to-roll manufacturing packaging	Low-cost packaging simplified production process	Large high-efficiency production unit
R&D	Large area deposition process	Improved deposition technique	New device structure with novel concepts

Source IEA technology Roadmap, 2010

1.5 Materials, Cells and Modules

Silicon continues to be the basic material used for the production of PV modules. Of all the semiconductor materials, the electrical, optical and physical properties of silicon have been most rigorously researched. Silicon-based PV cells have proven to offer good reliability in outer space as well as terrestrial applications. Other materials, such as gallium arsenide (GaAs) and cadmium telluride (CdTe), are more expensive. A disadvantage of silicon is that it is expensive to purify.

1.6 Physical Characteristics

Solar cells for terrestrial applications are typically made from silicon as single-crystal, polycrystalline or amorphous solids.

- Single-crystal silicon is the most efficient because the crystal is free of grain boundaries, which are defects in the crystal structure caused by variations in the lattice that tend to decrease the electrical and thermal conductivity of the material. They can be thought of as barriers to electron flow.
- Polycrystalline silicon has obvious grain boundaries; the portions of single crystals are visible to the naked eye.
- Amorphous silicon (a-Si) is the non-crystalline form of silicon where the atoms are arranged in a relatively haphazard way. Due to the disordered nature of the material, some atoms have a dangling bond that disrupts the flow of electrons. Amorphous silicon has lowest power conversion efficiencies of the three types, but is the least expensive to produce.

1.6.1 Single Crystal or Monocrystalline Silicon Cell

The work horse of PV industry has always been the crystalline silicon cell, fabricated from a single crystal or cast polycrystalline silicon that is sliced into wafer of about 10 × 10 cm area and 350 μm thickness as shown in Fig. 8.

These cells show efficiencies between 13 and 15 % depending on the material quality and the specific cell technology.

The single crystal of silicon is defined as having a grain size greater than 10 cm. Modules made of this type of cell are the most mature on the market. Reliable manufacturers of this type of PV module offer guarantees of up to 20–25 years at 80 % of nameplate rating.

1.6.2 Polycrystalline or Multicrystalline Silicon

These cells are made up of various silicon crystals formed from an ingot. They are also sliced and then doped and etched. They demonstrate conversion efficiencies slightly lower than those of monocrystalline cells, generally from 13 to 15 %. Reliable manufacturers typically guarantee polycrystalline PV modules for 20 years. Multicrystalline and polycrystalline cells respectively have grain sizes which are 1 μm–1 mm and 1 mm–10 cm. Nanocrystalline cells have grain size of less than 100 nm. Figure 9 shows an example of polycrystalline cell.

Today, the vast majority of PV modules (85–90 % of the global annual market) are based on wafer-based c-Si. Crystalline silicon PV modules are expected to remain a dominant PV technology until at least 2020, with a forecasted market share of about 50 % by that time. This is due to their proven and reliable

Fig. 8 A monocrystalline silicon cell

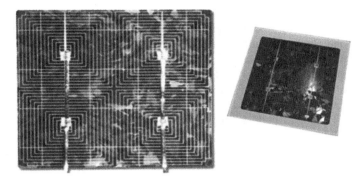

Fig. 9 A polycrystalline silicon cell (Lund et al., 2008)

technology, long lifetimes, and abundant primary resources. The main challenge for c-Si modules is to improve the efficiency (see Table 2) and effectiveness of resource consumption through materials reduction, improved cell concepts and automation of manufacturing (IEA International Energy Agency 2013).

1.7 Semiconductors

Semiconductors such as gallium arsenide (GaAs), aluminium gallium arsenide (AlGaAs), indium antimonide (InSb) and indium phosphide (InP) have considerable interest owing to their good characteristics. Most of these solar cells offer high efficiencies under increased irradiance and hence are ideally suited for concentrating devices. On the loss side, the optical losses accounted for up to 10 % of the incident energy. Further, concentrators do not effectively focus diffuse solar energy, and therefore their applications are restricted to sunny locations.

Concentrators approaching 40 % efficiency have been investigated: a multi-junction and tandem arrangement has been used. In a tandem arrangement two or more cells are stacked such that light not absorbed but passing through the top cell is utilized by the bottom device.

1.8 Thin Film (Non-silicon)

The high cost of crystalline silicon solar cell is primarily for its high cost of Si wafers, which constitute 40–50 % of the cost of the finished modules. Therefore, thin-film technology came into picture with the objective of reducing the cost of the material by using lesser quantity of absorbing material. The selected materials are all strong light absorbers and only need to be about 1 micron thick. Thus material costs are significantly reduced. Some of the most attractive materials used for thin-film solar cells are a-Si: H, cadmium telluride (CdTe), copper indium diselenide (CIS) and copper indium gallium diselenide (CIGS).

Thin-film modules are constructed by depositing extremely thin layers of photosensitive materials onto a low-cost backing such as glass, stainless steel or plastic. Thin-film solar cells consist of layers of active materials about 10 nm thick. The individual cells deposited next to each other, instead of being mechanically assembled. In order to build up a practically useful voltage from thin-film cells, their manufacture usually includes a laser scribing sequence that enables the front and back of adjacent cells to be directly interconnected in series with no need for further solder connection between cells.

However, commercially available thin-film PV has not attained field efficiencies greater than 10 %, as compared with the 16–18 % efficiency of crystalline silicon PV modules. Though laboratory tests have yielded promising thin-film efficiencies, manufacturers have not yet translated the high efficiencies and high yields of smaller, laboratory-constructed thin films up to production volumes.

Leading contenders are copper indium diselenide ($CuInSe_2$), copper indium sulphide ($CuInS_2$), copper indium telluride ($CuInTe_2$), cadmium sulphide (CdS), and cadmium telluride (CdTe).

The main advantages of thin films are their relatively low consumption of raw materials, high automation and production efficiency, ease of building integration and improved appearance, good performance at high ambient temperature, and reduced sensitivity to overheating. The current drawbacks are lower efficiency and the industry's limited experience with lifetime performances.

1.9 Amorphous Thin Film

The first thin-film solar cell produced was amorphous silicon (a-Si). Amorphous thin films have evolved from an efficiency of 2–5 % to above 12 %. Stability concerns have, however, been reported for this technology. Changes in performance after exposure to light are well known and the reported efficiencies seem to be those measured before any light-induced changes occurred. Figure 10 shows thin-film PV modules.

Based on early a-Si single junction cells, amorphous tandem and triple cell configuration have been developed. To reach higher efficiencies, thin amorphous and microcrystalline silicon cells have been combined to form micromorph cells (also called thin hybrid silicon cells). Another option currently being researched is the combination of single-crystalline and amorphous PV cell technology. The HIT (heterojunction with intrinsic thin layer cells) technology is based on a crystalline silicon cell coated with a supplementary amorphous PV cell to increase the efficiency.

Fig. 10 Thin-film PV
modules

1.10 Emerging Solar PV Technology

Emerging Solar Photovoltaic technologies, such as organic PV cells and
dye-sensitized solar cells are still under demonstration and have not yet been
commercially deployed on a large scale. They are also called third-generation solar
PV technology and have been described below:

1. Dye-sensitized solar cells use photoelectrochemical solar cells, which are based
   on semiconductor structures formed between a photosensitised anode and an
   electrolyte. In a typical DSSC, the semiconductor nanocrystals serve as antennae
   that harvest the sunlight (photons) and the dye molecule is responsible for the
   charge separation (photocurrent). It is unique in that it mimics natural photo-
   synthesis. These cells are attractive because they use low-cost materials and are
   simple to manufacture. They release electrons from, for example titanium
   dioxide covered by a light absorbing pigment. However, their performance can
   degrade over time with exposure to UV light and the use of a liquid electrolyte
   can be problematic when there is a risk of freezing. Laboratory efficiencies of
   around 15 % have been achieved due to the development of new broadband
   dyes and electrolytes, however, commercial efficiencies are low—typically
   under 4–5 %. The main reason why efficiencies of DSSC are low is because
   there are very few dyes that can absorb a broad spectral range. An interesting
   area of research is the use of nanocrystalline semiconductors that can allow
   DSSCs to have a broad spectral coverage. Thousands of organic dyes have been
   studied and tested in order to design, synthesize and assemble nanostructured
   materials that will allow higher power conversion efficiencies for DSSCs. DSSC
   are currently the most efficient third generation solar technology available.
2. Organic solar cells are composed of organic or polymer materials (such as
   organic polymers or small organic molecules). They are inexpensive, but not
   very efficient. They are emerging as a niche technology, but their future

development is not clear. Their success in recent years has been due to many significant improvements that have led to higher efficiencies. Organic PV module efficiencies are now in the range 4–5 % for commercial systems and 6–8 % in the laboratory. In addition to the low efficiency, a major challenge for organic solar cells is their instability over time. Organic cell production uses high-speed and low-temperature roll-to-roll manufacturing processes and standard printing technologies.

As a result, organic solar cells may be able to compete with other PV technologies in some applications, because manufacturing costs are continuing to decline and are expected to reach USD 0.50/W by 2020. Organic cells can be applied to plastic sheets in a manner similar to the printing and coating industries, meaning that organic solar cells are lightweight and flexible, making them ideal for mobile phones, laptops, radios, flashlights, toys and almost any hand-held device that uses a battery. The modules can be fixed almost anywhere to anything, or they can be incorporated into the housing of a device. They can also be rolled up or folded for storage when not in use. These properties will make organic PV modules attractive for building-integrated applications as it will expand the range of shapes and forms where PV systems can be applied. Another advantage is that the technology uses abundant non-toxic materials and is based on a very scalable process.

1.11 Efficiency

The performance of a solar cell is measured in terms of its conversion efficiency at converting sunlight into electricity, i.e. the efficiency of a PV device is defined in terms of the power produced from the incident photons. The common approach is to obtain the light-generated current–voltage (I–V) characteristics and ascertain the maximum power point. The standard measurement and reporting conditions are defined by reference to spectral irradiance, total irradiance, temperature and area of the device.

The following standard test conditions are used when quoting cell efficiencies for terrestrial applications: global irradiance = 1000 W/m^2, AM (air-mass) 1.5 and a temperature of 25 °C. The entire frontal area of the PV cell, including the grid, is taken into account when measuring its performance. The module takes into account the total area including the frame.

A typical commercial solar cell has an efficiency of 15 %—about one-sixth of the sunlight striking the cell generates electricity. Improving solar cell efficiencies while holding down the cost per cell is an important goal of the PV industry.

The following graph (Fig. 11) shows the best research cell efficiency of different PV technologies published by NREL National Center for Photovoltaics (2014).

As shown in Fig. 11, First Solar, Inc. set a world record for a cadmium telluride module conversion efficiency of 16.1 percent. This means that thin-film solar panels, which are much cheaper to produce, are getting more efficient.

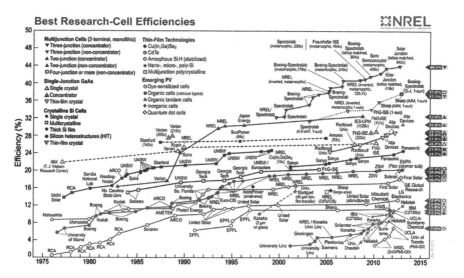

Fig. 11 NREL chart of record solar cell performance

1.12   PV Modules

A PV module or panel is a grouping of PV cells. The voltage generated by a single PV cell is inconveniently low. Several cells are always joined in series so that their voltages add up to a more useful value. The series connection of cells forms a unit called a PV module. The front consists of a window of low-iron glass with high transmission characteristic, which protects the surface of the PV material. Figure 12 below shows the connection of PV cells that form a module.

1.13   PV String

A string is a group of modules which are wired in series. This is to increase the voltage as modern solar electric DC systems operate minimum at 48 volts nominal, and for high-voltage grid-tied systems produce up to 600 volts. Figure 13 shows an example of a PV string.

1.14   PV Array

One of the attractions of PV modules is that connecting modules to form array may increase the power rating of a system. Within a PV array, there are basically two methods of connection between modules.

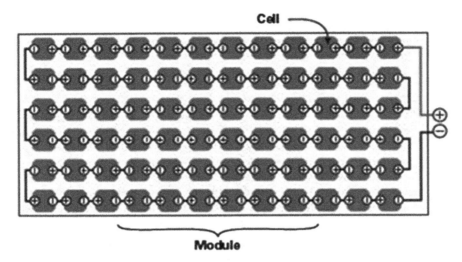

Fig. 12 Connection of PV cells that form a module

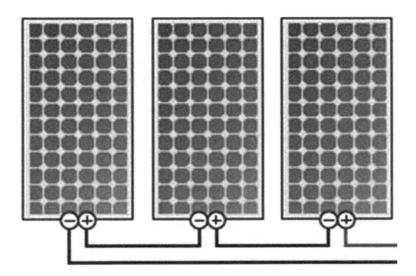

Fig. 13 An example of a PV string (Mason et al. 2008)

To increase the array voltage, modules are connected in series, known as series string. To increase the generating capacity without changing the array voltage, strings of modules are connected in parallel. Figure 14 shows a PV array connected in a series string.

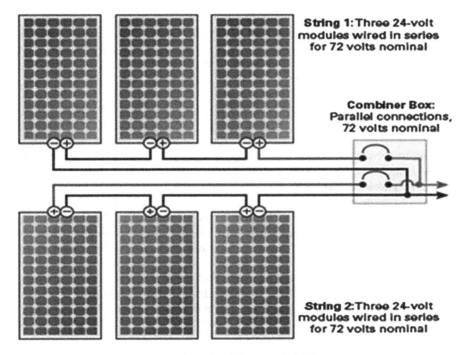

Fig. 14 PV array connected in a series string (Mason et al. 2008)

1.15 Size and Shape

PV cells generally come in the form of rectangular panels of variable sizes, or in the form of individual solar slates, that are installed on the roof in very much the same way as roof tiles. Both the rectangular panels, and the solar slates, are invariably fitted in multiples and they can be fitted in any configuration. For example, they can be fitted in portrait format, landscape format, in the shape of a T (to avoid roof lights), or in a random format.

The solar slates are always integrated into the roof but the rectangular panels can either be integrated into the roof or fitted on mounting brackets above the roof. The chosen installation method has no effect on the output of the panels. In general PV systems in buildings are sized in such a way that the PV system can meet the building loads either fully or partially and still function reliably. In stand-alone and hybrid systems, the batteries and/or backup system (i.e. diesel generator) must deliver the electricity during long overcast periods. In grid-connected systems, there is no storage component because the grid acts as a buffer.

The key factors affecting the system sizing are the load size, the operation time, the location of the system and a possible sizing safety margin. Besides that, the available roof or facade area can restrict the PV array size. Finally, the most important restriction for PV system sizing is the available budget. Roof/facade area and budget are typically the key restrictions for the design of a grid-connected PV house.

1.16 Structure Support and Fixing

Modules may be mounted on a pole, a ground support, a wall of a building, a building or vehicle roof or on a boat deck. The main considerations are day-long access to unobstructed sunlight and cable lengths to batteries.

Ideally, panels should be placed so that they are perpendicular to the noon-day sun, i.e. facing the equator and at an angle of inclination approximately equal to the angle of latitude, although they will function when mounted flat. A steeper angle of inclination will enhance output during winter months when the sun is lower in the sky at the expense of some reduced output in summer. Mounting frames can easily be fabricated. To avoid electrolytic corrosion aluminium angle and plated or stainless steel nuts and bolts should be used. Furthermore, mounts should be strong and capable of withstanding wind. Alternatively, a range of high-quality standard kits are available. These are pre-drilled to accept modules and include stainless steel fasteners.

1.17 Comparative Analysis of Different PV Technologies

Table 4 compares the PV technologies and PV devices. The comparison includes the efficiencies of the various PV modules (at an air mass index of 1.5), the area required to produce 1 kW of electricity, the stage of commercialization, the advantages and the drawbacks of the respective technologies.

1.18 Trend in Batteries

The majority of current energy generation is based on fossil fuels. There are a series of issues related to fossil fuels: continuous increase in demand for coal, oil and natural gas, depletion of non-renewable resources and dependency on politically unstable oil producing countries. Another worrying concern about fossil fuels is the CO_2 emission, which has almost doubled since 1970, resulting in rise in global temperature. Hence, it is essential that the present energy generation should be replaced urgently by clean energy sources. The CO_2 issue and consequent air pollution can be globally controlled by the deployment of mature renewable energy resources like wind and solar power plants. However, the intermittence of these resources requires high efficiency energy storage systems. Electrochemical systems

Table 4 Comparison between PV technologies and PV devices

Technology	Photovoltaic Device	PV Module Efficiency at AM 1.5 (%)	Area Req'd/kW (m²)	State of Commercialization	Advantages	Drawbacks
Wafer-based Silicon	Monocrystalline	15–19	7	Mature with large scale production	Suitable for applications in low to warm temperature areas and where there are space constraints	Operates at decreased efficiencies in higher temperature Less shadow tolerance
	Polycrystalline	13–15	8	Mature with large scale production	Lower manufacturing cost as compared to monocrystalline Higher shadow tolerance than monocrystalline	Less space efficient than mono-crystalline Less durable than mono-crystalline Less temperature tolerance than mono-crystalline
Thin-film cells	Amorphous silicon	5–8	15	Early deployment phase with medium scale production	High shadow tolerance Low manufacturing cost Serial connection of cells can be done during manufacture a-Si cells dominate the small output market (clocks, calculators, etc.)	Strong performance degradation under light exposure Low space efficiency
	Cadmium telluride	8–11	11	Early deployment phase with small scale production	Lowest specific cost (cost/watt) Low degradation rate ($\sim$ 0.9 %/yr) Serial connection is easy and in an integral manner over a large substrate	Cd and Te are toxic materials, although the compound is stable and harmless Low availability of Te
	CIGS	7–11	10	Early deployment phase with medium scale production	Flexible production technology. High scope of improvement in future	High manufacturing cost compared to the efficiency obtained Low availability of Indium

(continued)

Table 4 (continued)

Technology	Photovoltaic Device	PV Module Efficiency at AM 1.5 (%)	Area Req'd/kW (m²)	State of Commercialization	Advantages	Drawbacks
New emerging technologies	CPV	25–30	–	Just commercialized, small scale production	Highest efficiency achieved in PV technology	High manufacturing cost Requires two axis tracking, which further adds to the cost.
	Organic PV	1 %	–	R&D stage	Infinite variability of starting materials and low cost manufacturing processes Energy payback time is expected to be short	Long term stability and protection against environmental influences are significant challenge
	Dye sensitized solar cell	1–5 %	–	R&D stage	Less stringent purity requirement of starting material and a simple processing without clean room step is possible as surface/bulk recombination cannot happen	Low photocurrent efficiency with respect to incident light. Decreased stability with time and over different temperature ranges.

such as batteries can efficiently store and deliver energy demand in stand-alone power plants, as well as provide power quality and load levelling of the electrical grid in integrated systems (Scrosati and Garche 2010). For more than a century, people have been familiar with battery technology. This section presents analysis of the trend of battery technology which has become a key component for very many applications.

In 1801, the first battery was invented by Alessandro Volta by alternating one upon the other stacked Copper and Zinc (Cu/Zn) plates and these plates were separated by cloths, which were soaked in acids. Table 5 describes the brief historical overview of battery technology (Birke et al. 2010).

Of all battery technologies, lead acid batteries have had the longest developmental history. They came into market in early nineteenth century and due to its advantages like low manufacturing cost, good performance and long life, it still holds 40–45 % of the market. Similarly, Nickel cadmium batteries are mature and thoroughly tested having been patented in 1899. It is used in a wide variety of stationary and portable applications. However, due to stricter European environmental legislation, these batteries are expected to be gradually phased out from Europe. Further, the nickel metal hydride battery technology was developed in the early 1990s which offered the same cell voltage as nickel cadmium batteries, and was used to replace them in many applications without any modifications. Cell voltage combined with higher energy density and environmental properties are the major driving forces that enabled the present market share capture of nickel cadmium batteries.

As compared to metal hydride batteries, currently lithium-ion batteries are also widely used and have more benefits: high energy density levels, relatively high

Table 5 Historical overview of batteries

Year	Technology
1802	Ritter's column (first secondary battery) In 1802 he built the first accumulator with 50 copper discs separated by cardboard discs moistened by a salt solution
1859	First lead acid (Pb) battery
1899	First nickel cadmium (NiCd) battery (pocket plates)
1901	Nickel iron battery
1950s	Serial production of sealed nickel cadmium production
1972	Development of NaS (sodium–sulphur batteries) high temperature batteries
1980	CSIR Laboratory development of NaNiCl (sodium nickel chloride) ZEBRA battery
1983	Lithium metal rechargeable
1990	Introduction of nickel metal hydride (NiMH) batteries (Sanyo) with higher energy density and banned cadmium
1991	Introduction of lithium-ion batteries (Sony): Cobalt based
1996	Manganese-based lithium-ion batteries—cost optimized
1999	Lithium ion polymer
2002	Introduction of NMC cathode material
2004	Introduction LiFePO4 cathode material

Table 6 Li-Ion battery characteristics, Serra (2012)

Factor	NCA	LFP	LMO	LTO
Specific energy (Wh/kg)	170	140	150	150
Maturity	Most proven	Electronic monitoring need proving	Safety and durability needs proving	Safety and durability needs proving
Life expectance	Good	Moderate	High	High
Safety	Least thermal stability, high charge thermal runway	Least risk of overcharge	Better than NCA, some thermal instability expected	Better than LMO and NCA
Cost	High	Low	High	Highest

voltages and a low weight to volume ratio. However, still metal hydride batteries are used in some applications or as a low cost alternative to lithium ion. In practical terms, the term Li–ion covers a broad range of chemical composition. The four main types are (a) lithium nickel cobalt aluminium (NCA), (b) lithium iron phosphate (LFP), (c) lithium manganese polymer (LMO) and (d) lithium titanate (LTO). Table 6 provides details of all four types of Li-ion batteries.

Besides Li-ion batteries, another rising technology in market is high temperature batteries, i.e. sodium nickle chloride (NaNiCl) and sodium sulphur (NaS). The difference between high temperature batteries and other commercial batteries like lead–acid is that high temperature batteries have liquid electrode and a ceramic solid state electrolyte. High temperature is necessary to keep electrode in a molten state and to achieve sufficient ion conductivity in the electrolyte. Only molten electrodes can take part in charge or discharge reaction.

Currently, NaS batteries are only used in stationary application such as load levelling of electricity supply. In this case the battery is charged during the night-time to store cheap energy and discharged during the daytime peak load period. Similarly, NaNiCl2 can also be used in stationary applications but some developers use in electric vehicles as well.

In general, a battery is composed of several electrochemical cells stacked together. A typical cell configuration has three primary elements: a negative electrode (anode), a positive electrode (cathode) and both of them are immersed in an electrolyte (electrically conductive) substance. Hence, the battery-based energy storage systems are burdened by hefty battery packing and its overall weight challenge. Research done by Ehsani et al. (2004) shows that 26 % of a battery pack's overall weight is directly associated with the energy production process. It is mainly due to a need of sealed casing to prevent harmful chemicals from leaking into environment and to the wiring and electrical accessories. Hence, most of the current research is linked with specific energy of the battery. Specific energy is also known as gravimetric energy density. It is used to define the amount of energy the battery can store per unit mass. It is usually expressed in Wh/kg (Garcia-Valle and

Table 7 Comparison of energy densities by battery types

Battery type	Theoretical specific energy (Wh/kg)	Practical specific energy (Wh/kg)
Lead acid (Pb)	161	20–40
Nickel cadmium (NiCd)	240	25–45
Nickel zinc (NiZn)	320	45–80
Nickel metal hydride (NiMh)	300	45–80
Zinc bromine (ZnBr)	435	50–90
Sodium nickle chloride (NaNiCl)	720	80–110
Sodium sulphur (NaS)	795	90–140
Lithium ion (Li-Ion)	450	70–200

Peças-Lopes 2013). Table 7 indicates the theoretical and practical specific energy of various types of batteries.

For the deployment of renewable resources it is essential that the battery cost should be as low as possible and hence understanding the battery cost breakdown is important to assess the potential to reduce the overall battery cost. In terms of battery cost breakdown, there are three main categories which require consideration: materials, manufacturing and other items (corporate overhead, research and development, marketing, transportation, warranty cost and profit). Each of these categories is disaggregated into cell, module and pack level integration. A single cell is a complete battery with two current leads and separate compartment holding electrode, separator and electrolyte. A module is composed of a few cells either by physical attachments or by welding in between cells. A pack of batteries is composed of modules and placed in single containing for thermal management. Table 8 shows the battery cost breakdown of Li-ion batteries on the basis of cell, module and pack level (Anderson 2009).

Though the cost mentioned in Table 4 is reliable, there are several major problems associated with the estimation of battery cost. It is often difficult to determine exactly what the battery cost comprises, i.e. is it the cost of battery cells or battery pack? Does it include the cost of the full energy management systems? Is it a cost target set by the manufacturer? Is it a cost of unit purchase or large batch? Such questions need to be addressed before considering the battery cost. Table 9 illustrates the historical per energy cost of Li–ion batteries and energy density.

Table 8 Cost breakdown of Li-ion battery ($/kWh)

Level of integration	Cost category			Total ($/kWh)
	Materials	Manufacturing	Other	
Cell	734.53	23.15	86.90	844.59
Module	771.79	26.77	86.90	885.47
Pack	864.38	31.68	230.27	1126.33

Table 9 Yearly comparisons of battery cost and energy density, Whitmore (2014)

Year	Battery cost ($/kWh)	Energy density (Wh/L)
2008	1000	590
2009	900	700
2010	800	800
2011	620	1100
2012	500	1390
2013	300	1500

Energy density is the volumetric energy density, i.e. nominal battery energy per unit volume. It is often expressed as Wh/L.

In present market, Li–ion batteries are widely considered as one of the most promising technologies. However, for decades, many researchers are working to improve its performance (Lache et al. 2008). Some of the research aspects include (a) transition to cheaper and less toxic electrode materials (cathodes) including phosphate, and silicates, (b) transition to materials that have higher, reversible lithium reception, i.e. greater absorption of lithium leads to higher battery capacity (c) the development of materials that can withstand rapid charges (d) batteries for automotive and stationary applications, i.e. power supplies and energy supplies (d) Increased cell size in a form of stored energy capacity (e) battery system with high voltage level, including electrolyte that can withstand higher electrode potential without degrading or reacting with environment (f) battery system with enhanced safety as compared to current battery types.

Lastly, it can be concluded that Li–ion technology has not yet reached to its full potential. It started competing in the market since 2012 and there are many improvements needed. However, for low end applications, Lead acid or Nickel–Zinc will still be interesting options and according to researchers (Pillot 2014), lead acid battery will be at first position till 2020 in terms of volume and cost.

1.19   Trend in Inverters

The inverter is an essential component of a PV system as it is responsible for the effective conversion of the variable DC output of the PV modules into clean, sinusoidal AC current with the required frequency of either 50 Hz or 60 Hz. The optimal sizing of an inverter is dependent on the output from the PV generator which is controlled by the local climate, the surface orientation and inclination of the modules. PV modules have negative temperature coefficients of power that affect the power output which is dependent on solar radiation, cell temperature and the solar spectrum. The cell temperature, however, rises as the intensity of irradiance increases thereby limiting the optimum performance of the modules below the nominal DC power.

A reduction in the intensity of the radiation that reaches the surface of the module reduces its output power below its rated capacity thereby causing the inverter to operate under part load conditions with reduced system efficiency if the inverter is not properly sized. For optimum performance of a PV system, the rated capacity of the inverter should be higher than the rated capacity of the system in order to prevent operations at overload conditions. Improper sizing of inverters thus increases the total energy costs.

The selection of an inverter is dependent on its optimal features such as efficiency, frequency regulation, low standby losses, power correction factor, ease of servicing, cost and reliability. The inverter's modularity facilitates continuous scaling of its capacity to optimize energy for all conditions including lower irradiance conditions. One distinct feature of the high yield inverter is its ability to automatically reconfigure itself for continuous power generation in the event of any module inverter trip. It also has inbuilt mechanisms that rotate the standby and active duty of each module in order to share generating capacity thereby extending the service life of the inverter. At lower solar irradiation level, the inverter automatically resizes its modular solution to match the power output of the modules. A picture of an inverter is shown in Fig. 15 below.

Various types of inverter are used in solar PV applications.These are given below:

1.19.1 Single Stage/Central Inverter

The single-stage inverter (central inverter) is widely used for large scale power applications. Here, the single power processing stage takes care of all the tasks of maximum power point tracking (MPPT), voltage amplification and grid-side current control. In this configuration, the solar modules are connected in series to create strings with output voltage high enough to avoid an additional voltage boost stage. In order to obtain the desired power level, the strings are connected in parallel through interconnection diodes (string diodes) as shown in Fig. 16.

Although this configuration is widely used, the global efficiency of the generation system is effectively reduced. The main reason of reduced performance is due to the centralized MPPT control that fixes a common operating point for all PV modules (shaded as well as unshaded) whereas different operating point should be

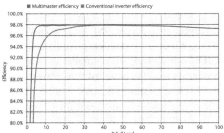

Fig. 15 An 'Emerson' inverter

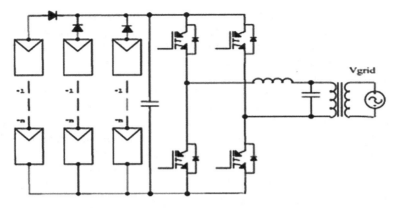

Fig. 16 Connections of solar PV strings

adopted for each module in order to extract the maximum power from the source. Because of these limitations, more advanced inverter topology is used based on the use of PV fields arranged in strings rather than arrays.

1.19.2 Double or Multi-stage Inverter

Here each string is connected to a double- or a single-stage inverter. If a large number of modules are connected in series to obtain an open circuit voltage higher than 360 V, the DC/DC converter can be eliminated. On the other hand, if a few number of PV modules are connected in series; a DC/DC boost converter is used. The DC–DC converter is responsible for the MPPT and the DC–AC inverter controls the grid current.

1.19.3 Multi-string Multi-stage Inverters with High Frequency Transformer

Another topology adopted is multi-string; multi-stage inverter. The multi-string inverter has been developed to combine the advantage of higher energy yield of a string inverter with the lower costs of a central inverter. Lower power DC/DC converters are connected to individual PV strings. Each PV string has its own MPPT, which independently optimizes the energy output from each PV string. All DC/DC converters are connected via a DC bus through a central inverter to the grid. Depending on the size of the string the input voltage ranges between 125 and 750 V. Here system efficiency is higher due to the application of MPPT control on each string and higher flexibility comes from the ease of extensions for the photovoltaic field. This topology is more convenient for power levels below 10 kW. The multi-string inverters provide a very wide input voltage range (due to the additional DC/DC-stage) which gives the user better freedom in the design of the

PV system. However, the disadvantages are that it requires two power conversion stages to allow individual tracking at the inputs.

1.19.4   Inverter Conversion Efficiency

Figure 17 presents the performance of a range of inverters.

Figure 17 shows that the efficiency range for all the inverters varies from 94.5 to 98.7 %. In the medium scale range (10–20 kW), there are several inverters available with the European (euro) efficiency range of 97–98 % and the incremental cost of these inverters is not much than that of the low efficiency inverter (generally USD 50/kW).

The 'European Efficiency' is an averaged operating efficiency over a yearly power distribution corresponding to middle-Europe climate. This was proposed by the Joint Research Center (JRC/Ispra), based on the Ispra climate (Italy), and is now referenced on almost any inverter datasheet.

Thus the project designer can evaluate the cost versus benefit of the inverter in terms of the enhanced efficiency.

1.19.5   Charge Controllers

The charge controller is an essential component in any PV/battery system as it prevents the batteries from damage that may arise from being overly discharged or overcharged. It controls the flow of current to and from the batteries once the maximum or minimum point of charge and discharge has been reached. Since most batteries hardly recover after exceeding the maximum depth of discharge, the integration of controllers in PV systems prevents this occurrence thereby extending the service life of the batteries. There are different types of charge controllers with

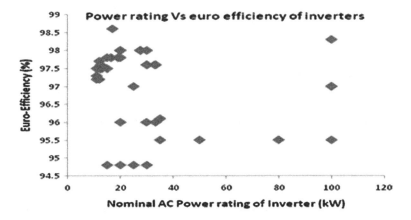

Fig. 17 Power rating versus euro efficiency of inverters

different operation mechanisms namely shunt controls, single-stage controls, multi-stage controls and pulse controls.

Shunt controllers are designed and utilised in small systems as they primarily prevent the flow of current that may result to overcharging by alienating the batteries once they have been fully charged. This is achieved by monitoring the maximum point of charge for the batteries and converting the excess power into heat. This heat is then dissipated by the shunt controllers which have heat sinks that require adequate ventilation for cooling.

With single-stage controllers, the batteries are protected once they are fully charged by switching off the current. The predetermined full state of charge is the charge termination set point (CTSP). At any point in the system when the battery power is drained up to the minimum discharge set point, the single state controller reconnects the power source to enable charging. This point is the change resumption set point (CRSP). The utilization of sensors by the single state controllers for the prevention of reverse current flow helps in the reduction of the heat produced as they break the circuit thereby eradicating the need for ventilation.

In PV/battery systems that utilize multi-stage controllers, the batteries are charged automatically based on their state of charge at any time. The flow of current is permitted once the batteries are at a low charge state. Dissipation of the array power occurs when the batteries approach full charge state. This mechanism helps prolong the service life of the batteries. Multi-stage controllers like the shunt controllers, also require ventilation as heat is generated during the dissipation of power.

Controllers implement effective load management by using low voltage disconnect (LVD). For the design of critical loads, warning lights are required for the indication of the battery's state of charge. The sizing of a controller requires parity with the systems' voltage. This is in addition to the capability of the controller to handle the flow of maximum PV current. Since the capacity and service life of batteries are affected by ambient and operating temperatures, the choice of controllers with temperature compensation features is very essential in the design of a system.

1.20 Cabling

Power losses in direct current (DC) systems are due to voltage drop. Care must be taken when selecting the size of the cable to be used. If a small cable is used then the voltage drop increases and in an off grid or stand-alone system this can have a major impact on the system where the battery voltage is lower than expected. Voltage drop can be calculated as follows:

$$V = I \times R$$

where V is the volt drop in the cable, I the current in the cable and R the resistance of the cable in ohms. Note that resistance depends on the length and cross sectional area of the cable. Table 10 shows the electrical resistance of the cable .

Conductor cross sectional area (mm²)	Resistance (Ω/m)
2.5	0.0074
4	0.0046
6	0.0031
10	0.0018
16	0.0012
25	0.00073
35	0.00049

Table 10 Electrical resistance of cable

Example 1 Calculate the voltage drop in a 100-m cable that has a 10 mm² cross-sectional area, carrying a current of 20A.

Solution:

$$\text{Voltage drop} = 20 \times 0.0018 \times 100 = 3.6$$

1.21 Metering

PV systems connected to the national grid are called **grid-connected** systems. This system provides benefits to the owner to get credit for the electrical energy produced by the PV system. Normally two metres are used in the grid-connected systems. One records the amount of energy produced by the PV system and the other; the amount of energy supply by the grid. Note that in some installations, a single metre is used—it goes backwards when local power is being generated, and forwards when power is being consumed. When installing a grid connected system, it is up to the local electricity authority as to which configuration of meters they approve.

2 Trend in Solar Radiation Measurement and Modelling Techniques

2.1 Introduction to Solar Radiation

Solar irradiation availability of arbitrary sloped surfaces is a prerequisite in many sciences. For example, agricultural meteorology, photobiology, animal husbandry, daylighting, comfort air-conditioning, building sciences and solar energy utilisation, all require insolation availability on slopes.

The past three decades have seen a boom in the construction of energy efficient buildings which use solar architectural features to maximize the exploitation of daylight, solar heat and solar-driven ventilation and solar PV electricity.

The initial research related to solar radiation carried out by Angstrom and others was concerned with the relationship between irradiation and the sunshine duration. Since then research in this field has come a long way. Today, a considerable amount of information is available on mathematical models that relate solar radiation to other meteorological parameters such as temperature, cloud cover, rain amount, humidity and even visibility.

2.2 Fundamentals of Solar Radiation

In this section algorithms are presented which enable calculation of the sun's position and the related geometry. The present set of algorithms includes low to high accuracy models. The high precision algorithm for solar position calculation was developed by Yallop (1992), a leading astronomer. Disparate practice has been adopted by meteorological stations across the globe in measuring hourly solar radiation. While in the UK the irradiation data is available against apparent solar time (AST), many other countries use the local civil time (LCT) as the reference for all records. Under the CIE IDMP the illuminance was recorded worldwide against the LCT. It is therefore necessary that appropriate algorithms are available for the conversion from one system to another. Basic concepts and definitions are introduced herein, which are a prerequisite for obtaining sun's position.

2.2.1 Day Number

In many solar energy applications one needs to calculate the day number (DN) corresponding to a given date. DN is defined as the number of days elapsed in a given year up to a particular date. Examples of this application are the estimation of the equation of time (EOT) and the solar declination angle (DEC) using low precision algorithms, and the extraterrestrial irradiance and illuminance at any given time.

2.2.2 Julian Day Number and Day of the Week

In many astronomical calculations it is often necessary to count the number of days elapsed since a predetermined reference date (fundamental epoch). By convention this date has been fixed as the Greenwich mean noon of 1 January 4713 BC. The number of days elapsed from this epoch to any given date is called the Julian day number (JDN). The calculation for JDN involves six steps and these are given in (Duffett-Smith 1988). The estimation of the day of the week is easy once JDN is known. This involves a three step algorithm which is also available in Duffett-Smith. Modern control algorithms for energy efficient buildings, such as optimum start algorithm, may find the use for the 'day of the week' routine to differentiate between weekdays and weekends.

2.2.3 Equation of Time

The difference between the standard time and solar time is defined as the EOT. EOT may be obtained as expressed by Woolf (1968):

$$EOT = 0.1236\sin x - 0.0043\cos x + 0.1538\sin 2x + 0.0608\cos 2x \quad (1)$$

where x = 360(DN − 1)/365.242, DN = 1 for 1 January in any given year. EOT may also be obtained more precisely as presented by Lamm (1981) as:

$$EOT = \sum_{k=0}^{5} A_k \cos(2\pi KN/365.25) + B_k \sin(2\pi kN/365.25) \quad (2)$$

where N is the day in the 4-year cycle starting after the leap year. Values of the A_k and B_k coefficients are given in Table 11. In any non-leap year, EOT assumes the value of near zero for 0 h UT for 15 April, 13 June, 1 September and 25 December.

2.2.4 Apparent Solar Time

Solar time is the time to be used in all solar geometry calculations. It is necessary to apply the corrections due to the difference between the longitude of the given locality (LONG) and the longitude of the standard time meridian (LSM). This correction is needed in addition to the above-mentioned EOT. Thus:

$$AST = \text{standard time}(LCT) + EOT \pm [(LSM - LONG)/15] \quad (3)$$

All terms in the above equation are to be expressed in hours. The algebraic sign preceding the longitudinal correction terms contained in the square brackets should

Table 11 Coefficients for Eq. (2)

k	$A_k \times 10^3$ (h)	$B_k \times 10^3$ (h)
0	0.2087	0.00000
1	9.2869	−122.29000
2	−52.2580	−156.98000
3	−1.3077	−5.16020
4	−2.1867	−2.98230
5	−1.5100	−0.23463

be inserted as positive for longitudes which lie east of LSM and vice versa. The LSM and LONG themselves have no sign associated with them.

2.2.5 Solar Declination

The angle between the earth–sun vector and the equatorial plane is called the DEC. As an adopted convention DEC is considered to be positive when the earth–sun vector lies northwards of the equatorial plane. Declination may also be defined as the angular position of the sun at noon (AST) with respect to the equatorial plane.

DEC may be obtained as expressed by Boes and reported in Kreider and Kreith (1981):

$$DEC = \sin^{-1}\{0.39795cos[0.98563(DN - 173)]\} \qquad (4)$$

Note that in the above equation, the cosine term is to be expressed in degrees. The arc sine term will obviously be returned in radian.

2.2.6 Solar Geometry, SOLALT and SOLAZM

The sun's position in the sky can be described in terms of two angles: SOLALT, the elevation angle above the horizon and SOLAZM, the azimuth from north of the sun's beam projection on the horizontal plane (clockwise is positive). These coordinates which describe the sun's position are dependent on GHA, the latitude (LAT) and longitude (LONG) of the location, and DEC. The solar geometry may now be obtained from the following equations:

$$\sin SOLALT = \sin LAT \sin DEC - \cos LAT \cos DEC \cos GHA \qquad (5)$$

$$\cos SOLAZM = \frac{\cos DEC \ (\cos LAT \tan DEC + \sin LAT \cos GHA)}{\cos SOLALT} \qquad (6)$$

2.3 Relevance of Solar Resources Assessment in Solar PV Plant Implementation

Solar resource is one of the most important inputs to PV power plant yield and performance evaluations. In order to assure well-founded decisions in designing profitable solar power plants, the solar irradiance should be measures in the assessment phase. Irradiation is a crucial parameter for site selection and plant design and economics of plant. There are many different ways and technologies to measure the irradiance phenomena that influences the power generation of a future solar power plant (Ammonit 2013).

2.4 Different Solar Radiation Measurement Techniques

Routine measurement of diffuse solar energy from sky and the global (total) radiation incident on a horizontal surface is usually undertaken by an agency such as the national meteorological office. For this purpose the measurement network uses pyranometers, solarimeters or actinograph. Figure 18 shows the picture of pyranometer.

Direct or beam irradiation is measured by a pyrheliometer with a fast-response multi-junction thermopile placed inside a narrow cavity tube. The aperture is designed such that it admits a cone of full angle around 6°. Most of the above irradiance sensors used across Europe are manufactured by Kipp and Zonen, while Eppley and Eko instruments are more widely used in the US and Japan, respectively. Table 12 summarised the characteristics of Kipp and Zonen's pyranometers.

The CM 22 is now regarded as the standard reference pyranometer due to its accuracy, stability and quality of construction. The sensing element consists of a thermal detector which responds to the total power absorbed without being selective to the spectral distribution of radiation. The heat energy generated by the absorption of radiation on the black disk flows through a thermal resistance to the heat sink. The resultant temperature difference across the thermal resistance of the disk is converted into a voltage which can be read by computer. The double glass construction minimizes temperature fluctuations from the natural elements and reduces thermal radiation losses to the atmosphere. The glass domes can collect debris over

Fig. 18 Pyranometers.
Source BADC (2014)

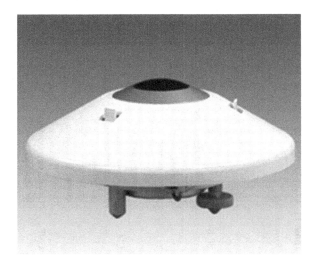

Table 12 Characteristics of Kipp and Zonen's pyranometers

Pyranometers		Spectral range	Sensitivity	Response time	Directional error (up to 80° with 1000 W/m² beam)	Temperature response	Operational temperature range	Maximum solar irradiance	Field of view
SP Lite2	Ideal for measuring available energy for use in solar energy applications, plant growth, thermal convection and evapotranspiration.	400–1100 nm	60–100 µV/W/m²	<500 ns	<10 W/m²	< −0.15 %/°C	−40 to +80 °C	2000 W/m²	180°
CM 4	Radiometer specially designed for measuring solar or artificial light irradiance under the most extreme temperature conditions.	300–2800 nm	4–10 µV/W/m²	< 8 s	< 20 W/m²	< 3 %	−40 to +150 °C	4000 W/m²	180°
CMP 3	Low cost pyranometer for accurate routine measurements in many applications.	300–2800 nm	5–20 µV/W/m²	18 s	< 20 W/m²	< 5 %	−40 to +80 °C	2000 W/m²	180°
SMP3	Perfect for monitoring solar energy installations, agriculture, horticulture, hydrological and industrial applications.	300–2800 nm	–	1.5–12 s	20 W/m²	< 3 %	–	–	–
CMP 6	For good quality measurements for green-house climate control, field testing and PV installations	285–2800 nm	5–20 µV/W/m²	18 s	20 W/m²	< 4 %	−40 to +80 °C	2000 W/m²	180°

(continued)

Table 12 (continued)

Pyranometers		Spectral range	Sensitivity	Response time	Directional error (up to 80° with 1000 W/m² beam)	Temperature response	Operational temperature range	Maximum solar irradiance	Field of view
CMP 10	Designed for meteorological networks and solar energy applications.	285–2800 nm	7–14 μV/W/m²	< 5 s	10 W/m²	< 1 %	−40 to +80 °C	4000 W/m²	180°
CMP 11	Reference measurements in extreme climates, polar or arid. It is the industry standard for solar radiation monitoring in PV and thermal energy plants.	285–2800 nm	7–14 μV/W/m²	< 5 s	10 W/m²	< 1 %	−40 to +80 °C	4000 W/m²	180°
SMP 11	Ideal choice for high quality solar radiation monitoring in meteorology and solar energy.	285–2800 nm	–	< 0.7 s– < 2 s	< 10 W/m²	< 1 %	–	–	–
CMP 21	Reference measurements in extreme climates, polar or arid.	285–2800 nm	7–14 μV/W/m²	5 s	< 10 W/m²	< 1 %	−40 to +80 °C	4000 W/m²	180°
CMP 22	For Scientific research requiring the highest level of measurement accuracy and reliability.	200–3600 nm	7–14 μV/W/m²	5 s	< 5 W/m²	< 0.5 %	−40 to +80 °C	4000 W/m²	180°

Source Kipp & Zonen (2014)

time and weekly cleaning is recommended. Moisture is prevented due to the presence of silica gel crystals in the body of the CM 11. The pyranometers have a spectral response of between 335 and 2200 nm of the solar spectrum which includes the visible wavelength band.

For most stations diffuse irradiance is measured by placing a shadow band over a pyranometer, adjustment of which is required periodically. Coulson (1975) provides an excellent account of these adjustments and the associated measurement errors for the above sensors, a brief summary of which is provided herein.

Radiation in the visible region of the spectrum is often evaluated with respect to its visual sensation effect on the human eye. The CIE meeting in 1924 resulted in the adoption of a standard of the above wavelength-dependent sensitivity. The CIE standardized sensitivity of daylight adapted human eye is presented in Table 13.

2.4.1 Equipment Error and Uncertainty

With any measurement there exist errors, some of which are systematic and others inherent of the equipment employed. Muneer has provided an account of the measurement errors associated with solar irradiance. These are summarised herein (Mohanty and Muneer 2004). The most common sources of error arise from the sensors and their construction. These are broken down into the most general types of error as follows:

a. cosine response
b. azimuth response
c. temperature response
d. spectral selectivity

Table 13 CIE standard spectral relative sensitivity of the daylight adapted

Wavelength (μm)	Relative sensitivity	Wavelength (μm)	Relative sensitivity	Wavelength (μm)	Relative sensitivity
0.38	0.0000	0.51	0.5030	0.64	0.1750
0.39	0.0001	0.52	0.7100	0.65	0.1070
0.40	0.0004	0.53	0.8620	0.66	0.0610
0.41	0.0012	0.54	0.9540	0.67	0.0320
0.42	0.0040	0.55	0.9950	0.68	0.0170
0.43	0.0116	0.56	0.9950	0.69	0.0082
0.44	0.0230	0.57	0.9520	0.70	0.0041
0.45	0.0380	0.58	0.8700	0.71	0.0021
0.46	0.0600	0.59	0.7570	0.72	0.0010
0.47	0.0910	0.60	0.6310	0.73	0.0005
0.48	0.1390	0.61	0.5030	0.74	0.0003
0.49	0.2080	0.62	0.3810	0.75	0.0001
0.50	0.3230	0.63	0.2650	0.76	0.0001

e. stability
f. non-linearity
g. thermal instability
h. zero offset due to nocturnal radiative cooling.

To be classed as a secondary standard instrument pyranometers have to meet the specifications set out by World Meteorological Organisation (WMO). Of all the aforementioned errors, the cosine effect is the most apparent and widely recognized. This is the sensor's response to the angle at which radiation strikes the sensing area. The more acute the angle of the sun, i.e. at sunrise and sunset, the greater this error (at altitude angles of sun below 6°). Cosine error is typically dealt with through the exclusion of the recorded data at sunrise and sunset times. The azimuth error is a result of imperfections of the glass domes, and in the case of solarimeters the angular reflection properties of the black paint. This is an inherent manufacturing error which yields a similar percentage error as the cosine effect. Like the azimuth error, the temperature response of the sensor is an individual fault for each cell. The photometers are thermostatically controlled, and hence the percentage error due to fluctuations in the sensor's temperature is reduced. The pyranometers rely on the two glass domes to prevent large temperature swings. Ventilation of the instrument is an additional recommended option. The spectral selectivity is dependent on the spectral absorptance of the black paint and the spectral transmission of the glass. The overall effect contributes only a small percentage error to the measurements. Each sensor possesses a high level of stability with the deterioration of the cells resulting in approximately ±1 % change in the full scale measurement per year. Finally, the non-linearity of the sensors is a concern especially with photometers. It is a function of irradiance levels. It however tends to contribute only a small percentage error towards the measured values. Table 14 provides details of the above-mentioned uncertainties. In addition to the above sources of

Table 14 WMO classification of pyranometers

Characteristic	Secondary standard	First class	Second class
Resolution (smallest detectable change in W/m²)	±1	±5	±10
Stability (percentage of full scale, change/year)	±1	±2	±5
Cosine response (percentage deviation from ideal at 10° solar elevation on a clear day)	<± 3	<± 7	<± 15
Azimuth response (percentage deviation from ideal at 10° solar elevation on a clear day)	<± 3	<± 5	<± 10
Temperature response (percentage maximum error due to change of ambient temperature within the operating range)	±1	±2	±5
Non-linearity (percentage of full scale)	±0.5	±2	±5
Spectral sensitivity (percentage deviation from mean absorptance 0.3–3 μm)	±2	±5	±10
Response time (99 % response)	<25 s	<1 min	<4 min

equipment-related errors care must be taken to avoid operational errors such as incorrect sensor levelling and orientation of the vertical sensors, as well as improper screening of the vertical sensors from ground-reflected radiation.

2.4.2 Types of Sensors and Their Accuracies

A survey of radiation instruments undertaken by Lof et al. showed that of the 219 sensors in use across Europe, 65 were of the CM11 type pyranometers while 107 sensors were the simpler and less expensive Robitzch actinographs with a bimet-tallic temperature element (Lof et al. 1965). The latter instrument is also quite popular in the developing Asian (89 such sensors were reported to be in use), African (16 sensors) and South American (47 sensors) countries where maintenance is often the key factor. The author has in the past visited a solar radiation mea-surement station in the middle of the Sahara desert and seen the Robitzch actino-graph faithfully recording a regular trace of irradiation. The weekly changeover of the recording chart makes this instrument an ideal choice for remote locations.

Drummond estimates that accuracies of 2–3 % are attainable for daily summa-tions of radiation for pyranometers of first class classification (Drummond 1965). Individual hourly summations even with carefully calibrated equipment may be in excess of 5 %. Coulson infers that the errors associated with routine observations may be well in excess of 10 %. Isolated cases of poorly maintained equipment but those which are in the regular network may exhibit monthly averaged errors of 10 % or more. The Robitzch actinograph, even with all the modifications to improve its accuracy is suitable only for daily summations. At this interval it provides an accuracy of around 10 %. However, not all designs of the latter sensor can claim even this level of accuracy. These figures must be borne in mind when evaluating the accuracy of the relevant computational models.

2.5   GIS Mapping of Solar Resource Potential

Developing solar radiation maps for a given region means creating illustrations revealing the geographical distribution of solar radiation covering that specific region. A solar radiation map demonstrates solar energy potentials of a specific region and provides information which is useful for optimum site selection of a solar energy system. A solar radiation map can be generated by using solar radi-ation data obtained from measurement stations. However, such a method is not applicable to many parts of the globe due to insufficiency of measurement stations. One solution is to use satellite-derived solar radiation data to create solar radiation maps (Gastli and Charabi 2010).

A Geographical Information System (GIS) is a system that can handle and process location and attribute data of spatial features (Kulkarni and Banerjee 2011).

GIS provides rapid, cost-efficient and accurate estimations of radiation over large territories, considering surface inclination, aspect and shadowing effects (Hofierka and Suri 2002). One of the first GIS-based solar radiation models was Solar Flux, developed for ARC/INFO GIS. Similar initiative was made by implementation of solar radiation algorithms into commercially available GIS Genasys using AML script.

More advanced methods for ecological and biological applications are used in Solar Analyst, developed as an ArcView GIS extension module. In the pre-processing phase, based on the DTM, the model generates an upward-looking hemispherical view shed. The similar procedure for generating a sun map for every raster cell makes calculation considerably faster. The model is suitable for detailed-scale studies. It is not flexible enough for calculation of atmospheric transmissivity and diffuse proportion as it allows to set parameters available only for the nearest weather stations or just typical values. This makes its use for larger areas rather limited. The SRAD model was designed to model a complex set of short-wave and long-wave interactions of solar energy with Earth surface and atmosphere. Although based on a simplified representation of the underlying physics, the main solar radiation factors are considered and the model is able to characterize the spatial variability of the landscape processes. However, it is designed for modelling of topo- and mesoscale processes and the calculation over large territories is also limited. A number of sources provide solar radiation data. Some of these are:

Furthermore note that 5 provides the details of designing and modelling procedure. Four case studies are also included within 5.

3   Conclusion

The global photovoltaic market in 2013 witnessed a massive growth worldwide with 38.4 GW of installation getting completed. During the period 2010–2014 the Crystalline Silicon PV module price dropped from 1.8–1 USD/W peak.

A review was presented of the three dominant PV technologies, i.e. crystalline, thin-film and emerging technologies. It was shown that the market share of crystalline PV dropped from 95 % in 2005 to 72 % in 2012. Over a longer term, its efficiency increased from 14 % in the year 1976 to 24 % in 2010, as reported by the US National Renewable Energy Laboratory. The present day efficiencies of some of the leading market contenders are: monocrystalline (15–19 %), polycrystalline (13–15 %), amorphous silicon (5–8 %), cadmium telluride and CIGS (7–11 %) and concentrating PV 25–30 %.

Energy storage will become increasingly important with the development of PV-based solutions and in this respect remarkable progress has taken place with lead–acid batteries providing storage capacities between 20 and 40 Wh/kg whereas lithium–ion technology introduced in 1991 is capable of providing 70–200Wh/kg capacity. Likewise the cost figures have also shown a dramatic improvement with a reduction from 1000 to 300 (USD/kWh taking place between the years 2008–13.

The volumetric energy density also increased from 590 to 1500 Wh/litre during the latter period.

Performance assessment of PV systems will require measurement of the solar resource and as such this chapter provided algorithms for obtaining solar geometry and also a detailed classification of pyranometers and their error and uncertainties.

Acknowledgments The authors of this chapter are grateful to Drs. Y. Tham and Y. Aldali for the help that they extended during the preparation of the initial draft.

References

Ammonit. Measurement GmbH. 2013 www.ammonit.com.

Anderson, D. (2009). An evaluation of current and future costs for lithium-ion batteries for use in electrified vehicle powertrains. Nicholas School of the Environment of Duke University.

BADC (British Atmospheric data Centre). Kipp & Zonen CG4 Pyrgemeter. Retrieved February 13, 2014, from https://badc.nerc.ac.uk/data/cardington/instr_v7/pyrgeometer.html.

Birke, P., Keller, M., & Schiemann, M. (2010). Electric Battery Actual and future Battery Technology Trends. Continental AG. Retrieved December 2, 2010, from http://www.futureage.eu/files/dd33e86df1_prezentace_Birke.pdf.

Coulson, K. L. (1975). *Solar and terrestrial radiation.* New York: Academic Press.

Drummond, A.J. (1965). Techniques for the measurement of solar and terrestrial radiation fluxes in plant biological research: a review with special reference to arid zones. In *Proceedings of the Montpiller Symposium,* UNESCO

Duffet-Smith, P. (1988). *Practical astronomy with your calculator* (3rd ed.). Cambridge, UK: Cambridge University Press.

Ehsani, M., Gao, Y., Gay, S.E., & Emadi, A. (2004). *Modern electric, hybrid electric and fuel cell vehicle: Fundamentals, theory and design.* Boca Raton, FL: CRC Press.

European Photovoltaic Industry Association (EPIA). Global market outlook for photovoltaics 2014–2018. Retrieved February 10, 2014, from http://www.epia.org/fileadmin/user_upload/Publications/44_epia_gmo_report_ver_17_mr.pdf.

Garcia-Valle, R., & Peças Lopes, J-A. (eds.). (2013). Electric vehicle integration into modern power networks, power electronics and power systems. New York: Springer Science Business Media New York. doi:10.1007/978-1-4614-0134-6-2.

Gastli, A., & Charabi, Y. (2010). Solar electricity prospects in Oman using GIS-based solar radiation maps. *Renewable and Sustainable Energy Reviews, 14,* 790–797.

Hofierka, J., & Súri, M. (2002). The solar radiation model for Open source GIS: Implementation and applications. In *Proceedings of the Open source GIS-GRASS Users Conference 2002,* Trento, Italy, 11–13 September 2002.

IEA International Energy Agency. (2013). Trends 2013 in photovoltaic applications. Survey Report of Selected IEA Countries between 1992 and 2012. Report IEA-PVPS T1-23:2013. ISBN 978-3-906042-14-5. Retrieved May 27, 2014, from http://www.iea-pvps.org/fileadmin/dam/public/report/statistics/FINAL_TRENDS_v1.02.pdf.

IEA Technology Roadmap, Solar photovoltaic energy. (2010). Retrieved February 13, 2015, from http://www.iea.org/publications/freepublications/publication/pv_roadmap.pdf. IRENA working paper. Renewable energy technologies: cost analysis series (Vol. 1). Power Sector. Issue 4/5. Solar Photovoltaics. June 2012.

Kipp & Zonen. (2014). Retrieved February 14, 2014, from http://www.kippzonen.com/.

Kreider, J. F., & Kreith, F. (1981). *Solar Energy Handbook.* New York: McGraw-Hill.

Kulkarni, S., & Banerjee, R. (2011). Renewable energy mapping in Maharashtra, India using GIS. In *World Renewable Energy Congress-Sweden*. 8–13 May, Linköping, Sweden. Sustainable Cities and Regions (SCR).

Lache, R., Galves, D., & Nolan, P. (2008). Electric cars: plugged in, batteries must be included, Deutsche Bank, June 9, Retrieved from Dutch INCERT website. Retrieved November 15, 2014, from http://www.d incert.nl/fileadmin/klanten/D-Incert/ webroot/Background_documents/DeutscheBank_Electric_Cars_Plugged_In_June2008.pdf.

Lamm, L. O. (1981). A new analytic expression for the equation of time. *Solar Energy, 26*, 465.

Lof, G. O. G., Duffie, J. A., & Smith, C. O. (1965). World distribution of solar radiation. *Solar Energy, 10*, 27.

Lund, H., Nilsen, R., Salomatova, O., Skåre, D., & Riisem, E. (2008). *The history highlight of solar sells [sic] (Photovoltaic Cells)*. Norwegian University of Science and Technology. http://org.ntnu.no/solarcells/pages/history.php.

Mason, J., Fthenakis, V., Zweibal, K., Hansen, T., & Nikolakakis, T. (2008). Coupling PV and CAES power plants to transform intermittent PV electricity into a dispatchable electricity source. *Progress in Photovoltaics: Research and Applications, 16*, 649–668.

Masson, G., Orlandi, S., & Rekinger, M. (2014). Global market outlook for photovoltaics 2014–2018. In Rowe, T. (ed.). Retrieved December 2, 2013, http://www.epia.org/fileadmin/user_upload/Publications/EPIA_Global_Market_Outlook_for_Photovoltaics_2014-2018_-_Medium_Res.pdf.

Mohanty, P., & Muneer, T. (2004). Smart design of stand-alone solar PV system for off-grid electrification projects. In S.C. Bhattaccharyya, & D. Palit (Eds.), *Mini-grids for rural electrification of developing countries*. Springer. doi:10.1007/978-3-319-04816-1.

NREL National Center for Photovoltaics. (2014). Research cell efficiency records. Retrieved April 15, 2014, from http://www.nrel.gov/ncpv/images/efficiency_chart.jpg.

Pillot, C. (2014). The worldwide battery market 2011–2025. Avicenne Energy. 2012. Retrieved November 27, 2014, from http://www.rechargebatteries.org/wp-content/uploads/2013/04/Batteries-2012-Avicenne-Energy-Batteries-Market-towards-2025l.pdf.

Reportlinker. (2011). Retrieved May 11, 2015, from http://www.reportlinker.com/p0233675/Thin—Film-Photovoltaic-PV-Cells-Market-Analysis-to-2020—CIGS-Copper-Indium-Gallium-Diselenide-to-Emerge-as-the-Major-Technology-by-2020.html?utm_source=prnewswire&utm_medium=pr&utm_campaign=Solar_Photovoltaic.

REN 21. (2014). Renewables 2014 Global Status Report. Paris: REN21 Secretariat. ISBN 978-3-9815934-2-6.

Scrosati, B., & Garche, J. (2010). Lithium batteries: Status, prospects and future. *Journal of Power Source, 195*(9), 2419–2430.

Serra, J. V. F. (2012). *Electric vehicles: Technology, policy and commercial development*. London: Earthscan. ISBN 978-1-84971-415-0.

Whitmore, A. (2014). Making a low carbon future better as well as cheaper. The energy collective. Retrieved December 1, 2014, http://theenergycollective.com/onclimatechangepolicy/347491/making-low-carbon-future-better-well-cheaper.

Woolf, H. M. (1968). *Report NASA TM-X-1646*. Moffet Field, CA: NASA.

PV System Design for Off-Grid Applications

Parimita Mohanty, K. Rahul Sharma, Mukesh Gujar, Mohan Kolhe
and Aimie Nazmin Azmi

Abstract Solar photovoltaic (PV) technology has the versatility and flexibility for developing off-grid electricity system for different regions, especially in remote rural areas. While conventionally straight forward designs were used to set up off-grid PV-based system in many areas for wide range of applications, it is now possible to adapt a smart design approach for the off-grid solar PV hybrid system. A range of off-grid system configurations are possible, depending upon load requirements and their electrical properties as well as on site-specific available energy resources. The overall goal of the off-gird system design should be such that it should provide maximum efficiency, reliability and flexibility at an affordable price. In this chapter, three basic PV systems, i.e. stand-alone, grid-connected and hybrid systems, are briefly described. These systems consider different load profiles and available solar radiations. A systematic approach has then been presented regarding sizing and designing of these systems. Guidelines for selection of PV components and system sizing are provided. Battery energy storage is the important component in the off-grid solar PV system. Due to load and PV output variations, battery energy storage is going to have frequent charging and discharging. So the type of battery used in a PV system is not the same as in an automobile application. Detailed guidelines for selection of battery are therefore also provided. At present, most of the world-wide PV systems are operating at maximum power points and not contributing effectively towards the energy management in the network. Unless properly managed and controlled, large-scale deployment of PV generators in off-grid system may create problems such as voltage fluctuations, frequency deviations, power quality problems

P. Mohanty (✉) · K.R. Sharma (✉) · M. Gujar
The Energy and Resources Institute, Lodhi Road, 110 003 Delhi, India
e-mail: parimatar.pm@gmail.com

M. Kolhe · A.N. Azmi
Faculty of Engineering and Science, University of Agder, PO Box 422 NO 4604,
Kristiansand, Norway

A.N. Azmi
Fakulti Kejuruteraan Elektrik, Universiti Teknikal Malaysia Melaka,
Durian Tunggal 76100, Malaysia

© Springer International Publishing Switzerland 2016
P. Mohanty et al. (eds.), *Solar Photovoltaic System Applications*,
Green Energy and Technology, DOI 10.1007/978-3-319-14663-8_3

in the network, changes in fault currents and protections settings, and congestion in the network. A possible solution to these problems is the concept of active generator. The active generator will be very flexible and able to manage the power delivery as in a conventional generator system. This active generator includes the PV array with combination of energy storage technologies with proper power conditioning devices. The PV array output is weather dependent, and therefore the PV power output predictability is important for operational planning of the off-grid system. Many manufacturers of PV system power condition devices are designing and developing new type of inverters, which can work for delivering the power from PV system in coordination with energy storage batteries as conventional power plant.

1 Introduction

This chapter is an introduction to guidelines and approaches followed for sizing and design of the off-grid stand-alone solar PV system. Generally, a range of off-grid system configurations are possible, from the more straightforward design to the relatively complex, depending upon its power requirements and load properties as well as site-specific available energy resources. However, the overall goal of the off-gird system design should be such that it can give maximum efficiency, reliability and flexibility of the system at an affordable price. While considering the above-mentioned points, the following sections cover the designing of solar PV system for off-grid electrification projects.

1.1 Types of Solar PV Systems

PV systems are broadly classified into three distinct types:

1. **Stand-alone systems** where the energy is generated and consumed in the same place and which does not interact with the main grid. Normally, the electricity consuming/utilizing device is part of the system, i.e. solar home systems, solar street lighting system, solar lanterns and solar power plants.
2. **Grid-connected systems** where the solar PV system is connected to the grid. The grid-connected system can either be a grid-tied system, which can only feed power into the grid and such system cannot deliver power locally during blackouts and emergencies because these systems have to be completely disconnected from the grid and have to be shut down as per national and international electrical safety standards. Some grid-connected PV systems with energy storage can also provide power locally in an islanding mode.
3. **Solar PV hybrid system**: In a hybrid system, another source(s) of energy, such as wind, biomass or diesel, can be hybridized with the solar PV system to

provide the required demand. In such type of system, main objective is to bring more reliability into the overall system at an affordable way by adding one or more energy source(s).

2 Guidelines for Designing of Stand-Alone Solar PV Systems

A systematic approach is important and required when sizing and designing stand-alone solar PV systems. The following procedures are generally followed:

A. *Planning and site survey;*
B. *Assessment of energy requirement;*
C. *Assessment of solar resource availability;*
D. *System concept development;*
E. *Sizing of main component of the PV systems; and*
F. *Selection of main components of the PV system.*

2.1 Planning and Site Survey

The PV array output depends on the geographical locations and timing. It is very important to select proper site based on solar resources. Therefore, in planning a PV system installation, appropriate selection of site with consideration of nearby high rise objects is necessary. The following points need to be covered during the site survey to check the suitability of the site (*Antony, Durschner, Remmers 2007*):

• site orientation, total land area/surface area of roof available;
• structure and type of roof; and
• possible routes for cables, battery and inverter location;

In order to identify the possible location for the installation of a solar PV module, probable options need to be assessed, such as "where exactly should the module be installed?" and "Can it be installed on the roof or on the ground?" (Fig. 1). If it is going to be installed on the roof, then whether the roof would be a thatched or a concrete or an asbestos one. And if it is going to be installed on the ground, then again the exact location needs to be identified.

The most critical parameter is the identification of a shadow-free location. It is to ensure that the solar PV array is installed in an area where no object casts a shadow on the array. For example, if there is a tall tree or a huge building in the vicinity of the selected location, then probably this is not a good location for the installation of the solar PV array. The required space needs to be identified appropriately, keeping in mind the capacity of the solar PV array, which is going to be installed in a particular location.

Fig. 1 PV system installation

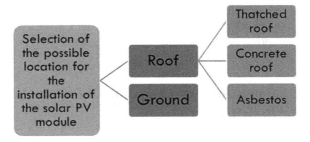

Hence, in order to address the above-mentioned points, the proposed site for installation can be surveyed with a solar path finder to check whether there is any possibility of seasonal shading problem. In addition to this, the most essential input parameters like incident solar radiation, the ambient temperature and wind velocity, which are likely to vary widely from site to site, need to be collected for a specific site. The air temperature and wind speed significantly affect the cell's operating temperature, and hence the output energy. So it is necessary to have access to necessary information on all these parameters in order to carry out an optimum PV system design exercise.

Box 1 Guidelines for PV system installation

Eliminating module shading by relocating the mounting system does not cost any additional money and can increase the system's efficiency by a large percentage. Inefficiency caused by excessive voltage drop in the system's wiring can also be inexpensively eliminated. Intelligent advance planning does not have high incremental cost, but can drastically reduce system's initial cost. In general, the designer should consider some important points while trying to optimize the system:

✓ Siting a system correctly so that it is not shaded;
✓ Orientation of the system is a critical element in maximizing annual PV output based on local meteorological conditions;
✓ Mounting options can maximize insolation gain;
✓ Modules should be selected according to a system's parameter;
✓ Wiring should be designed to minimize voltage drop;
✓ Controllers must operate system efficiently while meeting the needs of the system;
✓ Battery storage must be sized to a specific installation; and
✓ Loads determine the size of the system and should be scheduled by intelligent planning.

2.2   Assessment of Energy Requirement

In a stand-alone solar PV system, estimating the energy requirement and assessing the realistic solar resource availability are the most important tasks which have to be done properly. This is also critical from the point of view by adding smart load and resource management features. Following steps need to be considered for carrying out this exercise.

2.3   Load Assessment

In planning, seasonal and daily load variations are needed. It is important to assess the types and utilization of loads with their load profiles. An energy assessment should be undertaken for different types of appliances. A system designer needs to consider the energy requirements with load profiles in consultation with the consumers (Table 1). In process of load calculations, system designer should also discuss all the potential energy resources that can meet the energy needs of the consumer and also educate to the customers on energy efficiency.

Steps for load assessment

1. List all of the electrical appliances to be powered by the PV system.
2. Separate types of loads and enter them in the appropriate table.
3. Record the operating wattage of each item.
4. Specify the number of hours per day each item will be used.
5. Multiply step 2, 3 and 4 to calculate the daily energy requirement.

2.4   Load Profiling and Load Categorization

Once the load details are collected, the profiling of the load is done in order to find out the maximum load, average daily daytime energy requirement and average daily night time energy requirement. Loads can be categorized based on their priority and load profiles such as peak loads, off-peak loads and intermediate loads. In order to design a configuration and proper capacity of the energy storage system as well as the entire off-grid system, the loads should be considered based on their timing of operations. Furthermore, in order to give preference to certain loads over others in case of limited availability of energy, the loads can also be categorized based on their priorities, i.e. (i) critical or essential load and (ii) non-essential load. From the tariff payment point of view, the loads can also be categorized as based on operational requirements.

Table 1 Form for assessing the load and energy requirements

a	b	c	d	$e = c/d$	f	g	$h = e*g$	$i = c*f$
AC/DC Appliances/Load	Number	Power consumption (Wattage of each load) (W_{AC})	Power factor	Max Power demand (VA)	Daily usage hours (h/day)	Surge factor	Maximum surge demand (VA)	Daily energy requires (Wh_{AC}/day)
TV	–	–	–	–	–	–	–	–
Refrigerator	–	–	–	–	–	–	–	–
Lights	–	–	–	–	–	–	–	–
....	–	⌐	–	–	⌐	–	–	–

Total Wattage (of appliances) Sum of column c

Maximum demand
Sum of column e

Surge demand
Sum of column h

Total Daily AC energy required ($E_{AC\text{-}Daily}$)
Sum of column i

The information about the maximum demand and surge demand is collected in order to appropriately size and select an inverter for the solar PV system

2.5 Assessment of Solar Energy Resources

The assessment of solar energy resources is very important from the sizing point of view because it helps in estimating the output of the solar PV array. Incident solar radiation on PV array consists of direct radiation, diffuse radiation and ground reflected radiation. Solar radiation information is available through NASA by giving location latitude and longitude. While assessing the solar energy resource, the following information is important: average annual global solar radiation, average daily global solar radiation (direct and diffuse) on horizontal surface and no of sunshine hours in a year. A typical monthly averaged daily global and diffuse solar radiation on a horizontal surface is given in Fig. 2. However, since the power output of the PV array depends upon the incident solar radiation, the solar radiation falling on a horizontal surface needs to be adjusted with a tilt factor for calculating incident solar radiation on a tilted PV array.

2.6 System Configuration

An overall system configuration needs to be decided at stage of the design process. Several possible configurations need to be examined and the appropriate option should be selected at design stage considering energy demand, type of applications, spread of the community, willingness of the community, etc. In several cases, if the community has only lighting load requirement and the households are scattered,

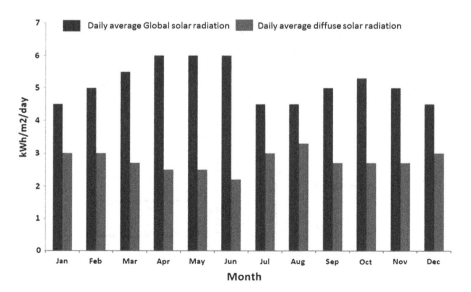

Fig. 2 Monthly averaged daily global and diffuse solar radiations on a horizontal surface

then probably Solar Home lighting System (SHS) or Solar Charging Station (SCS) will be preferred over solar mini-grid system. However, if the households are situated very closely and there is a demand for heavy motorized load, the centralized solar mini-grid may be preferred over SHS. Another thing that needs to be decided at this stage is the system voltage, especially in case of solar charging station, i.e. 12/24/48 V DC. A typical load variation with voltage is shown in Fig. 3.

- **Length of cable from solar PV module to battery**: Long DC distribution circuit may require higher system voltage in order to avoid heavy power loss or large cable diameter (if to keep the power loss under recommended limit).
- **Capacity or rating of the inverter** (if it is a large AC system): Generally, inverters over 2000 W are actually 24 V DC and inverters over 5000 W are often 48 V or above.

2.7   Sizing of Main Components of the PV System

The actual sizing of the PV system including its different components takes place, once the system configuration is decided. In general, a stand-alone solar PV system for off-grid applications majorly consists of (a) solar PV modules, (b) solar charge controller, (c) inverter, (d) storage batteries, (e) load and (f) other accessories such as cables, connectors, etc. Possible components, which are needed to consider in PV system design process, are given in Fig. 4.

The subsequent section explains the steps followed for sizing and designing of stand-alone solar PV power plant (without distribution network). The steps are based on a standard design procedures adopted universally. These steps can be customized further for designing of different configurations of PV system. For example, the steps associated with inverter in the standard sizing and design procedures can be omitted for a DC-based system without distribution network, whereas sizing of distribution network can be added to the standard procedures for AC-based mini-grid system.

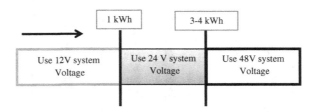

Fig. 3 A typical load variation with voltage

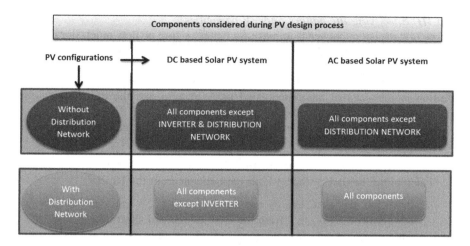

Fig. 4 PV system design process

2.8 Standard Sizing and Design Steps for Components of PV System

Step I: Assessment on energy requirement

The assessment on energy requirement (as explained in Sect. 3.2.2) is the first step of solar PV system sizing. Once this is carried out, the following steps are followed.

Step II: Specifying the inverter rating

An inverter is used in the system where AC power output is needed. The inverter must have the same nominal input voltage as the battery voltage. For stand-alone systems, the inverter must be large enough to handle the total amount of power that will be used at one time. The inverter size should be 20–25 % bigger than total power of appliances (as obtained from Table 1). Again the selected inverter should be capable of supplying continuous power to all AC loads and providing sufficient surge capability (Maximum demand and Surge demand as obtained from Table 1) to start any loads that may surge when turned on and particularly if they turn on at the same time.

Steps for determining inverter rating

1. Obtain "*Total power of AC appliances*", "*Maximum demand "and "Surge demand*" from Table 1.
2. Capacity or rating of the inverter should be 20–25 % bigger than "*Total power of AC appliances*".

   Capacity or rating of inverter ≥ 1. 2 X *Total power of AC appliances.*
   More details on selection of inverter are provided in Chapter 4.

Step III: Specifying the battery and PV capacities

The first decision that needs to make for battery sizing is 'how much storage you would like your battery bank to provide'. Often this is expressed as 'days of autonomy', because it is based on the number of days you expect your system to provide power without receiving an input charge from the solar PV array. In addition to the days of autonomy, load usage pattern and the criticality of the application should be considered.

While considering the battery sizing, the following parameters need to be considered:

i. **Daily Ampere-hour requirement:** The following steps can be adopted for calculating the daily Ampere-hour requirement from the battery.

  1. Enter the 'daily AC energy requirement $(E_{AC-daily})$' and 'daily DC energy requirement $(E_{DC-daily})$' from the load assessment sheet (Table 1).
  2. Enter the 'inverter efficiency' (η_{inv}) from the manufacturer's specification sheet.
  3. Divide the 'daily AC energy requirement $(E_{AC-daily})$' with inverter efficiency (η_{inv}) and add that to 'daily DC energy requirement $(E_{DC-Daily})$' to get the 'total daily DC energy requirement'.
  4. Enter the 'voltage of battery bank' (12 V/24 V/48 V or above).
  5. Divide the 'total daily DC energy requirement' with 'voltage of battery bank' to get the 'daily Ampere-hour requirement'.

ii. **Temperature effect**: Batteries are sensitive to temperature extremes, and you cannot take as much energy out of a cold battery as a warm one. Although you can get more than rated capacity from a hot battery, operation at hot temperatures will shorten battery life. Try to keep the batteries near room temperature. While sizing the battery, the winter ambient temperature multiplier (as shown in Table 2) should be considered in order to take care of the battery capacity in winter season.

iii. **Depth of Discharge**: The maximum depth of discharge value used for sizing should be the worst-case discharge that the battery will experience. The system control should be set to prevent discharge below this level.

iv. **Days of Autonomy**: to find out the days of autonomy, you need to take a decision on the following:

  (a) whether to increase the capacity of the solar PV array in such a way that there would not be any situation when load requirement is more than the generated PV energy and thus there is no extra day for storing energy in battery,
      or
  (b) certain number of extra days (generally 2–3 days) is considered for storing energy in battery,
      or

Table 2 Ambient temperature multiplier

Winter ambient temperature	Multiplier
80 °F (26.7 °C)	1.00
70 °F (21.2 °C)	1.04
60 °F (15.6 °C)	1.11
50 °F (10.0 °C)	1.19
40 °F (4.4 °C)	1.30
30 °F (−1.1 °C)	1.40
20 °F (−6.7 °C)	1.59

   (c) solar PV array is sized in such a way that it can cater the required energy in most of the months in a year, and in the remaining month, battery can be charged through another energy resource (such as diesel and/or wind generator) and thus there is no extra day (or maximum one day) for storing energy in battery.

The following worksheet (Table 3) is followed to carry out the sizing of the storage battery.

Table 3 Battery sizing worksheet

1. Enter your daily amp-hour (Ah) requirement	Ah/Day
2. Enter the maximum number of consecutive cloudy weather days expected in your area, or the number of days of autonomy you would like your system to support	_____
3. Multiply the amp-hour requirement by the number of days. This is the amount of amp-hours your system will need to store	Ah _____
4. Enter the depth of discharge[a] for the battery you have chosen. This provides a safety factor so that you can avoid over-draining your battery bank. (Generally minimum discharge limit is 20 % i.e. 0.2. and this should not exceed 0.8)	_____
5. Divide line 3 by line 4	Ah
6. Select the multiplier that corresponds to the average wintertime ambient temperature your battery bank will experience	_____
7. Multiply line 5 by line 6. This calculation ensures that your battery bank will have enough capacity to overcome cold weather effects. This number represents the total battery capacity you will need	Ah _____
8. Enter the amp-hour rating for the battery you have chosen (use the 20 or 24 h rate from the battery manufacturer)	_____
9. Divide the total battery capacity by the battery amp-hour rating and round off to the next highest number. This is the number of batteries wired in parallel required	_____
10. Divide the nominal system voltage (12 V/24 V/48 V) by the battery voltage and round off to the next highest number. This is the number of batteries wired in series	_____
11. Multiply line 9 by line 10. This is the total number of batteries required	_____

[a]The maximum allowable DoD of the battery depends upon the type as well as characteristic of the battery (*whether lead acid battery—flooded type, Sealed Maintenance Free (SMF), Valve Regulated Lead Acid (VRLA) or gel type; lithium (Li) based—lithium ion, lithium phosphate* etc.). More details about the selection of the batteries are given in Chap. 4

$$\text{Battery capacity (Ah)} = \frac{\text{Total Daily energy (Wh per day) required by appliances} \times \text{Days of autonomy}}{0.85 \times 0.6 \times \text{nominal battery voltage}}$$

$$(1)$$

Note (Box 2):
In a standard design, the battery capacity is done based on the total daily energy requirement, whereas in a smart design concept, the battery sizing is done based on the total nocturnal energy requirement and 5–8 % of the daytime energy requirement. In a standard design, the capacity of the battery is decided based on the total daily energy requirement. It does not segregate between the daytime energy requirement and nighttime energy requirement. However, the difference between these design approaches does not seem substantial if there is no or little daytime load and daytime energy requirement. But it makes considerable difference if there is considerable daytime energy requirement.

For a daytime load, there is no need of storing a large proportion of solar PV-generated DC electricity in battery. Rather, the DC electricity can be directly converted to AC electricity through the inverter. This would result not only improve the efficiency of the entire system (by around 5–7 %), but would also lead to smaller capacity of the battery and thus reduce the capital as well as replacement cost of the battery. However, the battery capacity should be selected that it can take care of the sudden fluctuation in the solar radiation (such as swift movement of clouds would result in a sudden drop of solar irradiation). So, in order to cater to the above points, on an average, 5–8 % of the daytime energy requirement can be added to the nocturnal energy requirement to find out the ideal daily energy requirement which is used to design the battery capacity:

Daily energy requirement (for battery sizing) E
$= $ nocturnal energy requirement $+ 0.08$ X day $-$ time energy requirement

$$(2)$$

Once the above-mentioned information is collected, all other steps mentioned above are followed to calculate the battery capacity. In order to determine the energy required from the PV array, it is necessary to increase the energy from the battery bank to account for battery efficiency. The average round-trip energy efficiency of a new battery is 80–85 % (variations in battery voltage are not considered). Therefore, the energy required, which needs to be provided by the solar PV array, is

Energy(Wh)to be provided by the solar PV array
= Daily energy requirement expressed in Wh ÷ 0.85   (3)

Oversize factor—If the system does not include a diesel generator which can provide extra charging to the battery bank, then the solar PV array should be oversized to provide the equalization charging of the battery bank. This is recommended as 10 %.

De-rating of module performance—Several factors (such as when you will be using your system—summer, winter, or year-round, location and angle of PV array, fixed mountings vs. trackers, etc.) influence how much solar insolation falls on the solar modules. Again, whatever solar insolation is falling on the solar module is not fully converted into useful electricity and there exist a power loss from solar PV module.

The PV array is de-rated due to various factors:

(a) Dirt and dust: Over a period of time, dirt or salt (if located near the coast) can build up on the array and reduce the output. The output of the module should therefore be de-rated to reflect this soiling. The actual value will be dependent on the site but this can vary from 0.9 to 1 (i.e. up to 10 % loss due to dirt).

(b) Temperature: Modules' output power decreases with temperature above 25 °C and increases with temperatures below 25 °C. The output power and/or current of the module must be based on the effective temperature of the cell.

The following worksheet (Table 4) is followed to carry out the sizing of the solar PV array.

Step IV : System wiring sizing

Cable or Conductor of working zone critical to the safe, long-term operation of any electrical system. This is particularly critical for PV applications, where the outdoor environment can be extreme and the PV modules will be sourcing current for 40 years or more. Cable sizes are particularly important for low voltage battery cables, solar panels and load cables. Voltage drops through incorrectly sized cables are one of the most common reasons for low voltage (12 V/24 V/48 V) system faults. If the cable is far too small, it can be very dangerous as the cable will heat up and potentially cause a fire. Undersized cables also waste energy.

A. Cable size between solar PV array and battery

The following steps can be followed to calculate the conductor size:

Step-1: Determine the maximum DC system voltage

In the DC side of the circuit, i.e. from the PV module side to the combiner box or to the inverter, calculate the maximum DC system voltage (shall not exceed the inverter maximum DC input voltage):

Maximum DC system voltage (volt) = Maximum number of modules per string
$$\times\ V_{oc} \times \text{temperature correction factor}$$

$$(4)$$

Table 4 Solar PV array sizing worksheet

1. Daily energy requirement	——Wh
2. Battery round-trip efficiency	80–85 %
3. Dividing the total energy demand per day (Line 1) by the battery round-trip efficiency (Line 2) determines the required PV array output per day	——Wh
4. Enter selected PV module maximum power voltage at standard test conditions (STC) (module specifications)	——V
5. Multiply 0.85 PV module maximum power voltage at STC to establish a design operating voltage for the solar module	—Vop
6. Enter nominal power output at 1000 watts/m^2 and 25^0 C (module specifications)	——W
7. Module de-rating factor	0.9
8. Multiply Line 6 with Line 7 to obtain the guaranteed power output	——W
9. Enter peak sun shine hour (Equivalent hours of Sun Shine—EHSS)	——h
10. Multiply Line 8 with Line 9 to get the average energy output from one module.	——Wh
11. Divide Line 3 with Line 10 to get the number of modules required to meet energy requirements	——Nos
12. Enter nominal power output of PV module	—W
13. Multiplying the number of modules to be purchased (Line 11) by the nominal rated module output (Line 12) determines the nominal rated array output	——
14. Enter the battery bus voltage	——V
15. Dividing the battery bus voltage (Line 14) by the module design operating voltage (line 5), and then rounding this figure to the next higher integer determines the number of modules required per string	—— Nos of module per string
16. Dividing the number of modules required to meet energy requirements (Line 11) by the number of modules required per string (Line 15) and then rounding this figure to the next higher integer determines the number of string in parallel	—— Nos of string in parallel

Step-2: Calculate the Maximum DC current

As per standard 690.8(A), in the DC PV array circuits, the maximum DC current is defined as 1.25 times the rated short-circuit current I_{sc} (module specification). For example, if a module had an I_{sc} of 7.5 amps, the maximum current would be $1.25 \times 7.5 = 9.4$ amps. If three strings of modules are connected in parallel, the PV output circuit of the combiner would have an I_{sc} of $3 \times 8.1 = 24.3$ amps. So the maximum current in this circuit would be $1.25 \times 24.3 = 30.4$ amps.

Step-3: Calculate the Maximum DC current to be carried by the conductor

For code calculations, PV currents are considered continuous and are based on worst-case outputs and based on safety factors applied to rated outputs. Because PV system currents are considered continuous, the maximum currents calculated as per standard 690.8(A) must be multiplied by 125 % to calculate the maximum continuous current or minimum conductor size/minimum ampacity:

Maximum continuous current to be carried by the conductor
$$= 1.25 \times \text{Maximum DC current} \qquad (5)$$

Hence,

Maximum continuous current to be carried by the conductor (ampacity)
$$= 1.25 \times 1.25 \times \text{rated short circuit current (Isc)} \qquad (6)$$

This calculation is done before applying any adjustment and correction factors, commonly referred to as 'conditions of use', which include corrections for conductors exposed to temperatures in excess of 30 °C or more than three current-carrying conductors within a conduit. The ampacity of the conductor, at a minimum, needs to be greater than or equal to the maximum current in given in standard 690.8(A) × 1.25.

Step-4: Enter percentage cable loss acceptable

Generally, the cable length and the cross-sectional area are chosen in a way that voltage drop between any two sections is within the permissible voltage level. Normally, 2–3 % voltage drop is allowed to calculate the cable length and the cross-sectional area.

Step-5: Calculate the cable length and the cross-sectional area

Once ampacity of the circuit is known, the maximum distance a cable can run in a 12-V DC system with expected maximum DC current, where 2 % voltage drop is allowed from PV array to charge controller or power converter can be found out using the chart as shown in Table 5 (source—http://www.affordable-solar.com/Learning-Center/Solar-Tools/wire-sizing).

The above chart gives the nominal ampacity, when ambient temperature is 30 °C. However, the ambient temperature and the cable operating temperature need not always be kept at 30 °C and thus the correction factor with respect to change in operating temperature needs to be considered to calculate the actual ampacity. The ambient temperature correction factor will depend upon

i. hottest outdoor temperature (expected) and
ii. the number of current-carrying conductors, running inside the conduit.

Table 6 (source—http://www.thesolarplanner.com/steps_page9b.html) shows the ambient temperature correction factor, whereas Table 7 (source—http://www.thesolarplanner.com/steps_page9b.html) shows adjustment factor with respect to the current-carrying conductors, running inside the conduit.

The actual ampacity after factoring the ambient temperature correction factor and adjustment factor with respect to the current-carrying conductors, running inside the conduit, will be

Actual ampacity = Nominal ampacity ÷ ambient temperature correction factor
          ÷ adjustment factor with respect to the current carrying conductors

$$(7)$$

Table 5 Voltage drop in cables

2 % Voltage drop chart										
	In sq mm									
	2.5	4	6	10	16	25	35	55	70	
	AWG									
Nominal Ampacity (Amps)	#14	#12	#10	#8	#6	#4	#2	#1/0	#2/0	#4/0
1	45	70	115	180	290	456	720	–	–	–
2	22.5	35	57.5	90	145	228	360	580	720	1060
4	10	17.5	27.5	45	72.5	114	180	290	360	580
6	7.5	12	17.5	30	47.5	75	120	193	243	380
8	5.5	8.5	11.5	22.5	35.5	57	90	145	180	290
10	4.5	7	11.5	18	28.5	45.5	72.5	115	145	230
15	3	4.5	7	12	19	30	48	76.5	96	150
20	2	3.5	5.5	9	14.5	22.5	36	57.5	72.5	116
25	1.8	2.8	4.5	7	11.5	18	29	46	58	92
30	1.5	2.4	3.5	6	9.5	15	24	38.5	48.5	77
40	–	–	2.8	4.5	7	11.5	18	29	36	56
50	–	–	2.3	3.6	5.5	9	14.5	23	29	46
100	–	–	–	–	2.9	4.6	7.2	11.5	14.5	23
150	–	–	–	–	–	–	4.8	7.7	9.7	15
200	–	–	–	–	–	–	3.6	5.8	7.3	11

Source http://www.affordable-solar.com/Learning-Center/Solar-Tools/wire-sizing

Table 6 Ambient temperature correction

Ambient temperature (°F)	Ambient temperature (°C)	Correction factor 75 °C conductors	Correction factor 90 °C conductors
70–77	21–25	1.05	1.04
78–86	26–30	1.00	1.00
87–95	31–35	0.94	0.96
96–104	36–40	0.88	0.91
105–113	41–45	0.82	0.87
114–122	46–50	0.75	0.82
123–131	51–55	0.67	0.76
132–140	56–60	0.58	0.71
141–158	61–70	0.33	0.58
159–176	71–80	0.00	0.41

Source http://www.thesolarplanner.com/steps_page9b.html

Table 7 Adjustment factor with respect to the current-carrying conductors	Number of current-carrying	Adjustment factor
	1–3 conductors	1.00
	4–6 conductors	0.80
	7–9 conductors	0.70
	10–20 conductors	0.50

Source http://www.thesolarplanner.com/steps_page9b.html

Once the actual ampacity is found out, it is used to find out the length and size of the cable which is to be used.

Alternatively, the cross-sectional area (*A*) of the cable is given by the equation

$$A = \frac{\rho l I}{Vd} \times 2 \qquad (8)$$

where ρ is the resistivity of copper wire which is 1.724×10^{-8} Ωm,

Vd is the maximum allowable voltage drop,

l is the length of the cable and

I is the maximum current that can be carried by the cable or the conductor.

Based on the above formula, the cross-sectional area (*A*) of the cable between PV module to charge controller, battery to the inverter or to the load, and between the inverter to the load can be calculated.

Determine the cable size between battery to load and inverter:

Let us consider the length of the cable (*l*) as 5 m and the allowable voltage drop is 4 %. In such scenario, the cross-sectional area is determined as follows.

The maximum current from battery at full load supply is given by

$$I = \frac{\text{Inverter } kVA}{\eta_{\text{inverter}} \times V_{\text{system}}} \qquad (9)$$

Here, V_{system} is the minimum possible voltage of the battery.

Since 4 % voltage drop is allowed, allowable maximum voltage drop (*Vd*) will be

$Vd = 0.04 \times 48\,\text{V} = 1.92\,\text{V}$ (Assuming that the system voltage is 48 V).

Applying the value of *l*, *Vd*, *I* and ρ in Eq. (8), the cross-sectional area of the cable can be calculated.

Determine the cable size between inverter and load:

Let us assume that the maximum length of the cable for powering the load from the inverter is 20 m and the allowable voltage drop is 4 %. So the maximum current on the phase is

$$I\text{max} = \frac{\text{Inverter } kVA}{\sqrt{3} \times V_{\text{output}}} \text{ (in case of 3 phase inverter)} \qquad (10)$$

Table 8 PV system component sizing

Component	Description of component	Result
Load	Total estimated load (kW) Total estimated energy(kWh)	
PV Array	Capacity of PV array	
Number of modules in series		
Number of modules in parallels		
Total number of modules		
Battery Bank	Battery bank capacity (Ah)	
Number of batteries in series		
Number of batteries in parallel		
Total number of batteries required		
Voltage Regulator	Capacity of voltage regulator	
Number of voltage regulators required		
Inverter	Capacity of the inverter (kW)	
Wire	Between PV modules and batteries through voltage regulators	
	Between battery bank and inverter	
	Between inverter and load	

$$I\text{max} = \frac{\text{Inverter} kVA}{V_{\text{output}}} \text{ (in case of single phase inverter)} \qquad (11)$$

The maximum continuous current will be $I = 1.25 \times I\text{max}$ and maximum allowable voltage drop (Vd) will be $Vd = 0.04 \times 220\,\text{V} = 8.8\,\text{V}$. Applying the values of l, Vd, I and ρ in Eq. 8, the cross-sectional area of the cable between inverter and load can be calculated. Once the sizing for different components of the PV system are done by following the above-mentioned steps, a summary table (Table 8) can be prepared.

3 Design and Actual Implementation of Solar PV System for Lighting and Livelihood Applications

In this section, design of various off-grid solar PV systems for lighting and livelihood generation activities will be described along with few examples of actual implementation of such systems.

3.1 Design and Actual Implementation of Solar PV System for Lighting Applications

Traditionally, solar lighting was provided through stand-alone individual systems such as solar lantern, Solar Home lighting System (SHS). However, in the recent years, in addition to the individual solar lighting systems, centralized solar systems too have come into use are also used for providing lighting options. Such centralized systems are available in the form of solar charging stations or small micro-grids or nano-grids. This entire configuration has its own advantages as well as limitations, depending on the requirement and applications. Whereas the sizing and design of the stand-alone solar lantern is very common and widely known, the authors exclude it in this book and focuses only on the sizing and actual implementation of solar charging station and solar micro-/nano-grid.

3.2 Design and Actual Implementation of Solar Charging Station for Lanterns

A Solar Charging Station (SCS) for lantern (Fig. 5) is a charging station where a number of lanterns charge simultaneously through a junction box (JB) using one or more solar PV modules that are centrally located. A solar lantern charging station facilitates the use of large capacity PV modules, which offer better efficiency and lowers unit costs as compared to the small capacity PV modules that are used individually and dedicatedly with a single solar lantern.

Following are its major components of the solar charging station to be used for charging lanterns or task lights.

PV modules—A set of PV modules is installed on the shadow-free area of the charging station. The voltage and current of each PV module are chosen in a way that it is capable of charging a particular pre-determined number of lanterns.

Lantern—A lantern is a portable lighting system consisting of lighting device (lamp), a maintenance-free storage battery and electronics that are all placed in a case made of plastic or fibreglass. During the day, the storage battery of the lanterns is charged through the JB ports by the electricity generated from the PV module. When the lantern is fully charged, it is disconnected from the JB and then can be used as an independent portable lighting source. Lantern is suitable for both indoor and outdoor lighting applications. The specifications of the lanterns are generally based on their light output and typical power rating.

Junction boxes—A JB basically contains the electronic interface circuitry that is required between the PV module and the lanterns. It houses the necessary protections such as short-circuit and reverse-polarity protections for effective charging of the lanterns. For proper distribution of current and the protection of the lanterns, the JB in SCS contains current limiting circuits for each individual port.

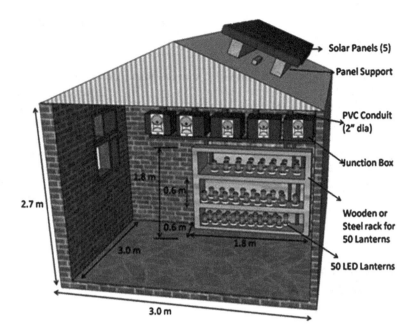

Fig. 5 A Solar Charging Station (SCS) for lantern. Source: TERI

Design of Solar Charging Station (SCS)—The concept of SCS has been tried out in many countries including India, Indonesia, Nepal and Myanmar as well as in many African countries. In those countries, SCS is implemented in order to charge lantern or task lights. The 'Lighting a Billion Lives (LaBL)' initiative taken by TERI had used this concept and so far implementing the largest number of SCS for charging lanterns (Fig. 6).

Step-1: Determine your power consumption and daily energy demand

Here, there is no AC loads and only have LED lights which are DC load. So the designer only has to use the format as per Table 9 to calculate the power consumption and daily energy demand. A typical example is shown in Table 9.

Step-2 Assess solar energy resources

In order to carry out this exercise, the location, where the SCS is going to be installed, has to be identified and the solar resource assessment needs to be carried out for that particular location. Let us assume that this particular sizing exercise is carried out for a SCS which is to be installed at New Delhi. The average daily insolation (sometimes referred to as "peak sun hours) for New Delhi is found out to be 4.8 kWh/m²/day, which is used in the sizing purposes in subsequent steps.

Fig. 6 A solar charging station implemented in India and Africa. *Source: TERI*

Table 9 Daily energy consumption of LED lighting

a	b	c	d	e	f = b*c*e
DC Appliances/Load	Number	Power consumption (Wattage of each load) (W_{DC})	Max Power (W max)	Daily usage hours (h/day)	Daily energy required (Wh $_{DC}$/day)
LED light	10	1.25	12.5	5	62.5
Maximum DC demand 15 W					
Total daily DC energy required ($E_{DC\text{-}Daily}$) 62.5 Wh					

Step-3 Specifying the inverter capacity

Since there is no AC load, there is no requirement of inverter.

Step-4: Specifying the battery

Here, the storage battery is with each lantern and thus the sizing of battery is carried out for each lantern. Here, it is assumed that small sealed maintenance-free (SMF) lead acid batteries are used in the lantern. The battery sizing worksheet (Table 10) is used to find out the battery capacity which is to be used for each lantern.

Step-5: Sizing of solar PV array

Although individual battery is used for each lantern, the sizing of solar module/array will be for simultaneously charging 10 numbers of lanterns. The solar module/array sizing worksheet (as mentioned in Table 11) is used to find out the solar PV capacity to be used for charging 10 lanterns.

Table 10 Battery sizing worksheet

1. Enter your daily amp-hour requirement for an individual lantern (Daily DC energy requirement/system voltage) (assuming that the driver efficiency is 0.95)	= 6.25 Wh/0.95/6 V = 1.1 Ah/Day
2. Enter the maximum number of consecutive cloudy weather days expected in your area, or the number of days of autonomy you would like your system to support (Days of Autonomy)	3 days
3. Multiply the amp-hour requirement by the number of days. This is the amount of amp-hours your system will need to store	= (1.1* 3)Ah = 3.3 Ah
4. Enter the depth of discharge for the battery you have chosen. This provides a safety factor so that you can avoid over-draining your battery bank	0.8
5. Divide line 3 by line 4	= (3.3/0.8) Ah = 4.1 Ah
6. Select the multiplier that corresponds to the average wintertime ambient temperature your battery bank will experience	1.1
7. Multiply line 5 by line 6. This calculation ensures that your battery bank will have enough capacity to overcome cold weather effects. This number represents the total battery capacity you will need	4.5 Ah
8. Enter the voltage and amp-hour rating for the battery you have chosen (use the 20 or 24 h rate from the battery manufacturer)	6 V, 4.5 Ah
9. Divide the total battery capacity by the battery amp-hour rating and round off to the next highest number. This is the number of batteries wired in parallel required	One
10. Divide the nominal system voltage (12, 24 or 48 V) by the battery voltage and round off to the next highest number. This is the number of batteries wired in series	One
11. Multiply line 9 by line 10. This is the total number of batteries required.	One 6 V, 4.5 Ah
Battery capacity (SMF lead acid battery) to be used for each lantern	6 V, 4.5 Ah

4 Role of Photovoltaic System in Future Smart Micro-Grid

4.1 What Is Future PV-Based Smart Micro-Grid Will Look Like?

Future electrical grid will be face numbers of physical changes with introduction of new technologies for integrating PV system. According to the report from MIT entitles 'the future of electrical grid', the challenge will be massive as large penetrations of intermittent resources into the grid system will change a lot of things in the grid system. This will significantly change the policy initiatives as well as design of the grid. A lot of countries around the world have put a renewable initiatives as part of their policy initiatives in order to face this huge challenge towards the conventional grid system which has been exist since early 1900. There

Table 11 PV module/array sizing worksheet

Daily energy requirement (for 10 lanterns)	62.5 Wh
Battery round-trip efficiency	80–85 %
Charging efficiency of the junction box	0.9
Dividing the total energy demand per day (Line 1) by the battery round-trip efficiency (Line 2) and charging efficiency of the junction box to determines the required array output per day	= (62.5/0.8/0.9) = 87 Wh
Enter selected PV module max power voltage at STC Maximum power voltage is obtained from the manufacturer's specifications for the selected photovoltaic module	8.3 V
Multiply 0.85 PV module max power voltage at STC to establish a design operating voltage for the solar module	= 8.3*0.85 Vop = 7 Vop
Enter nominal power output at 1000 W/m^2 and 25 °C	25 Wp
Module de-rating factor (including mismatch, soiling and other losses	0.9 (90 %)
Multiply line 6 with 7 to obtain the guaranteed power output at operating temperature	= 25*0.9 W = 22.5 Wp
Enter peak sun shine hour (Equivalent hours of Sun Shine (EHSS)	4.8 h
Multiply line 8 with 9 to get the average energy output from one module	= 108 Wh
Divide line 3 with 10 to get the number of modules required to meet energy requirements (No of modules purchased)	One
Enter nominal power output of PV module	25 Wp
Multiplying the number of modules to be purchased (line 11) by the nominal rated module output (line 12) determines the nominal rated array output	One 6 V, 25 Wp solar module

are needs to introduce a new type of smart grid system that really can plug and play rather than replacing all the conventional grids as it is not financial beneficial.

'Smart grid' is being publicized by most of the utility providers throughout the world. This can be seen as a future grid system. However, there are much more things that the power system needs to face to maintain it reliability and security. This is going to be one of the major engineering challenges ever. The transition from conventional to smart grid system can be represented in three stages.

The smart grid includes several crucial components (Fig. 7) and it will define a future grid system. As such system will turn into maturity after sometimes, it needs to be evolved according to the rapid technology growth. It is expected that the future grid will be a two-way flow of energy and information compared to current grid system which is unidirectional. Future grid is expected to be more resilient and capable of self-healing to any fault. Basic ingredients in intelligent grid can be seen in Table 12.

The modernization of power generation by utilization of technological advancement is to make sure a better efficiency, more reliable grid system and environmental friendly.

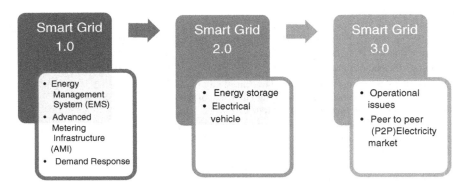

Fig. 7 Evolution of smart grid transition (Carvallo and Cooper 2011)

Table 12 Smart grid compared with existing grid (Farhangi 2010)

Existing grid	Intelligent grid
Electro-mechanical System	Digital system
One-way communication	Two-way communication
Centralized generation	Distributed generation
Hierarchical	Distributed network
Few sensors	Sensors throughout
Blind	Self-monitoring
Manual restorations	Self-healing
Failures and blackouts	Adaptive and islanding
Manual check/test	Remote check/test
Limited control	Pervasive control
Few customer choices	Many customer choices

4.2 Future Grid Interconnection Option for Solar PV Systems

This subsection will provide a future grid connection to the PV system with energy storage and it is going to be called an active generator. At present, most of the world-wide grid-connected PV systems are operating at maximum power points and not contributing effectively towards the energy management in the power system network. Unless properly managed and controlled, large-scale deployment of grid-connected PV generators may create problems: voltage fluctuations, frequency deviations and power quality problems in the power system network, changes in fault currents and protections settings and congestion in distributed network. These problems are becoming critical for maintaining the power system stability and control. A solution to these problems is the concept of active generator. The active generators will be very flexible and able to manage the power delivery as used to be in conventional generator system in micro-grid. This active generator includes the PV array with combination of energy storage technologies and proper power conditioning devices. Many manufacturers of PV system power condition devices are designing and developing new type of inverters, which can work for delivering the power from PV system in coordination with energy storage/batteries as conventional power plant.

The higher penetrations of distributed generators are going to create different possibilities of the faults not only in the micro-grid network but also at higher voltage power system network. Fault detection and isolation mechanism is compulsory for power system operation. It is needed to analyse the fault protection system, for instance: fault current levels, relay settings and fault clearing time in the micro-grid environment by considering the existence of PV-based active generators (Najy et al. 2013). During fault or any unwanted events and abnormal conditions at the micro-grid network, the grid may be disconnected, and islanding effect may occur in micro-grid. It will create many problems towards the grid operation and safety issues. In such situations, micro-grid EMS has to be intelligent for effectively managing the power flows within the micro-grid by considering not only voltage and frequency fluctuations but also taking into accounts the safety using different protection standards (Kunsman; Teodorescu and Liserre 2011). These standards are used to make sure that the PV-based active generator and grid connections are safe and not going to harm either equipment or personnel. Using these protection standards, the utilities company can envisage the impact of the control strategies of the connection, which includes the performance of voltage deviations, power quality and harmonics.

4.3 Overview of PV-Based Active Generator

PV-based active generator is a system that comprises PV array with a battery storage system with a capacity of storing energy for a long and short term for local usage (Kanchev et al. 2010). From this definition, it can be concluded that this system will be able to generate, store and release energy as long as the electricity is needed. This can be done with a proper hierarchical monitoring and energy management system. Figure 8 shows systematic of PV-based active generator.

Power management is crucial to control the whole energy flow in the PV-based active generator with the emphasis on the power management algorithm on the active PV station with a battery storage (Di Lu et al. 2008). Four hierarchical positions have been introduced and each level has its own task, as shown in Fig. 9. This PV-based active generator is expected to offer a new flexibility to the consumer and operator and will be a new dimension of generating electricity through clean energy. The system will be operated in a micro-grid environment and will have a lot more parameters that need to be considered.

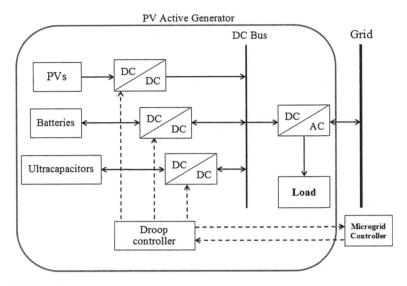

Fig. 8 PV-based active generator systematic

Fig. 9 PV generator control
level (Kanchev et al. 2011)

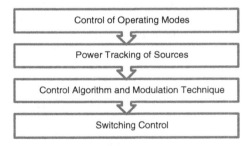

The system is connected to the battery storage and or ultra-capacitors and which is then coupled with choppers and connected to the existing grid (micro-grid). With the combination of battery and ultra-capacitors, it will increase the system efficiency as the battery will be able to store and release energy gradually, while ultra-capacitor effectively acts as storage device with very high power density. For a complete PV active-based generator, a set of battery bank is connected in a combination series–parallel in order to provide desired power to the system. The additional ultra-capacitor will provide a fast response energy storage device that can reduce the effect of short-term fluctuations of PV output and will enhance the whole system (Shah and Mithulananthan 2011).

4.4 Battery Storage and Ultra-Capacitor (Energy Storage System)

Battery storage sizing is very important. There are three main parameters that need to be considered for every installation of battery storage system: the depth of discharge (DOD), state of charge (SOC) and state of health (SOH), as well as battery capacity, maximum battery charge and discharge power and the utility rating type (Aichhorn et al. 2012). Battery storage systems are being progressively used in distributed renewable energy generation nowadays. With the existence of ultra-capacitors, the effectiveness of the storage system will be much more reliable to be used in the near future. With the combination of battery and ultra-capacitors, it will increase the system efficiency as the battery will be able to store and release energy gradually, while ultra-capacitor effectively acts as storage device with very high power density. For a complete PV active-based generator, a set of battery bank is connected in a combination series–parallel in order to provide desired power to the system. The additional ultra-capacitor will provide a fast response energy storage device that can reduce the effect of short-term fluctuations of PV output and will enhance the whole system. Basically, the total produced power from the system is a total power generated from the PV, battery and ultra-capacitor (Fig. 10).

A lot of research has been done for the battery storage system for PV generator. The battery dynamic equation can be represented as

$$\frac{dE_B}{dt} = P_B(t), \tag{12}$$

where E_B represents the amount of electricity stored at t time and P_B is the charging or discharging rate. This should be integrating with the ultra-capacitor to make sure that both of this storage system can be used and compatible to each other. There are several relevant resources regarding the optimization of batteries for PV. There are following levels of working zone and conditions for a stand-alone PV application, and they are

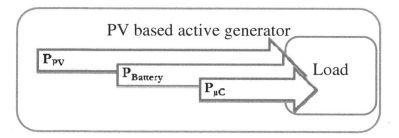

Fig. 10 Power flow in PV-based active generator

(i)   Saturation zone;
(ii)   Overcharge zone;
(iii) Charge zone;
(iv)   Changing from charge to discharge or vice versa;
(v)   Discharge zone; and
(vi)   Over discharge and
(vii) Exhaustion.

As can be seen in Fig. 11, the working condition depends on the voltage and current that went through the battery. This is a sample of a 2-V battery.

For basic facts on installation of storage system to a grid-connected PV system, there are three different facts (Vallvé et al. 2007). First, argument is the storage system that can undoubtedly improve the security of supply to the whole system; however, the grid quality is the main issue. The existing grid is ageing and the possibilities of interruption are high. Second, the addition of storage function might increase the performance ratio of PV generator. The third fact is that large penetration of PV will definitely not be able to cover the whole load consumptions. PV source can be able to supply at least some part of overall energy consumed by specific load. These facts may lead the utility company or customer to consider on the battery and ultra-capacitor as a main storage structure for future housing development. The Fraunhofer Institute for Solar Energy Systems has developed a battery-management system for renewable energy system. For a conventional renewable energy system, the battery often operated at the low state of charge that

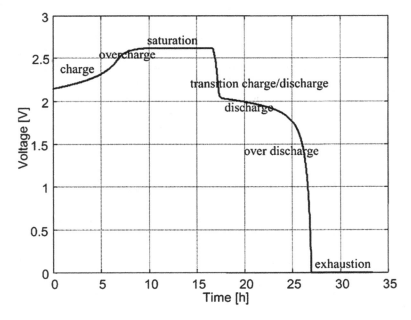

Fig. 11 Battery working zone conditions (Guasch and Silvestre 2003)

resulting the decreasing lifetime of the battery. By implementing a battery-management system on the renewable energy system, it improves the storage lifetime and reliability of batteries in the system and thus reduces maintenance and lifetime costs considerably.

The battery storage system for PV integration is also discussed further in (Le Dinh and Hayashi 2013), (Li et al. 2013), (Yoo et al. 2012). The authors in these papers discuss on the main topic related to connection of PV system to the micro-grid: frequency control, voltage stability and energy storage smoothing control. These parameters are significant for a PV-based active generator, since it is compulsory to get a very efficient system to make sure that it can be implemented in the real system. A smoothing control method for reducing output power fluctuations and regulating battery state of charge (SOC) under typical condition is proposed. Voltage control and active power control algorithm for centralized battery storage system is already proposed in (Le Dinh and Hayashi 2013).

There are a lot of new inventions on the energy storage system to fit in the new grid system or micro-grid with PV system going on. One of the discoveries is the utilization of Vanadium Redox Battery (VRB) in PV system. As discussed in (Wang et al. 2012b) and based on the evidence in (Nguyen et al. 2011) and (Wang et al. 2012a), the utilization of VRB in micro-grid has abundance of chance. Based on Table 13, the comparison between conventional Lithium ion (Li-ion), Sodium Sulphur Battery (NaS), Lead Acid battery, VRB and flow battery can be seen.

Table 13 Comparison between different types of batteries

Parameter/Technology	Li-ion	Na S	Lead acid	VRB	Flow battery
Energy density	Average	High	Low	Low	Varies (lower than Li-ion)
Efficiency	High (near 100 %)	High (~92 %)	85 %	~85 %	60–85 %
Lifecycle	500–1000		200–300	High	
Toxicity	Non-toxic (electrolyte may be harmful)	Highly Corrosive	Sulphuric acids in the lead is highly corrosive	No fire hazard / No highly reactive or toxic substances	Low toxicity
Cost	High (above ~ $600)	High (up scaling)	Low	High	Low on average (depends on type of chemical)
Other		High operation temperature (heating process needed)	Requires regular maintenance	Independent energy and power rating more complicated technology	High power and capacity for load levelling in grid system

For a PV system application, the battery storage system will be operated under the partial state of charge duty (Hill et al. 2012). In this condition, the battery or ultra-capacitors will be partially discharge at all time, in order to make sure that the system will be able to absorb or discharging power to the grid as it is needed (Guerrero et al. 2011). To charge the ultra-capacitor, a few methods can be done, and hence using a constant power charging mode will be better in PV environment; however, it is not proven that it will be suit to the micro-grid environment yet (Zhang et al. 2012). The charging efficiency using constant power charging mode is a ratio between energy in the ultra-capacitor (E_μ) and energy transmitted by the chargers (E_T):

$$\eta = \frac{E_\mu}{E_t} \tag{13}$$

And further it can be expressed by

$$\eta = \frac{\frac{1}{2} C_e (V_{cT}^2 - V_{c0}^2)}{PT} \tag{14}$$

where
C_e is the deal ultracapacitor value,
Vc_T is the ultracapacitor voltage as T time,
Vc_o is the ultracapacitor initial voltage and
P is the charging power.

For a dynamic equation of ultracapacitor, it can be seen as

$$\frac{C_e}{\omega} \frac{\partial V_c}{\partial t} = \frac{1}{R_e} \left[\left(\frac{1-d}{d} \right) V_{dc} - V_c \right] \tag{15}$$

where
C_e is the capacitance value,
R_e is the series resistance of ultracapacitor,
V_c is the ultracapacitor voltage and
d is the duty cycle.

The duty cycle implemented in the system is proportional and integration controller (PI).

4.5 PV-Based Active Generator in Future Micro-Grid Environment

Micro-grid is a system that operates at low voltage and has a few distributed energy resources (PV, wind, geothermal, etc.). With proper energy management and

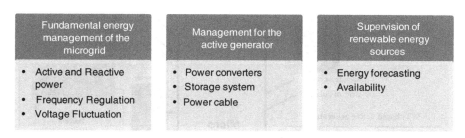

Fig. 12 Parameters that need to be considered in PV-based active generator energy management

systematic supervision, micro-grid can be a new dimension of generating and transmitting energy to the load. PV-based active generator can be integrated into micro-grid and it has been done in Kytnos Island in Greece, Tokoname city in Japan, Bronsbergen Holiday Park in The Netherlands and Mannheim-Wallstadt in Germany. It needs a good supervision from the utility operator to make sure that it will well operate. Energy supervision and optimization for the whole PV-based active generator system is compulsory and it can be separated into three major parts as shown in Fig. 12.

On the micro-grid side, the operator needs to manage the energy between source and load. This will includes the active and reactive power, frequency regulation, voltage fluctuations, etc. The implementation of PV-based active generator in a micro-grid environment has a high significant towards the energy management in the electrical grid. A strategic framework needs to be developed once it is executed in the system and it should be focused on a long-term and short-term energy management.

Figure 13 shows the micro-grid that integrates together three energy sources (PV, wind and micro-gas turbine). This is a representation of a micro-grid system with multiple sources that can be used in future micro-grid system.

4.6 Energy Management in Micro-Grid

For a better power delivery, the most crucial part is on the energy management side. Theoretically, it might look simple, yet it is tough. In micro-grid connection, other than findings, a new alternative optimization criteria and exploration of the fluctuations effect, energy management options, is important to be modelled so that a reliable energy with better efficiency can be delivered without any failure to customer (Quiggin et al. 2012). A deterministic energy management algorithm for a PV-based active generator in micro-grid environment needs to be set up for proper supervision.

Based on Fig. 8, the PV-based active generator will be coupled via a DC bus and will be connected to the micro-grid through a three-phase inverter. This will be connected and controlled by a micro-grid controller through a droop controller for

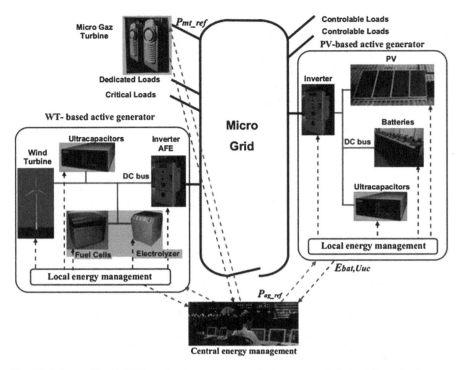

Fig. 13 Micro-grid with PV-based active generator and other sources (wind turbine and micro-gas turbine). Source: (D Lu and Francois 2009)

primary frequency control. A basic requirement for satisfactory operation of power system is the PV-based active generator which needs to maintain the nominal frequency of the grid (50 or 60 Hz). The rules of thumb for frequency control depend on active power (P), while voltage is based on reactive power (Q). Thus, for better energy management for PV-based active generator, a proper droop controller that will manage the voltage and frequency variation is a must.

For PV-based active generator, it will not engage any inertia of the mechanical system since there will be no kinetic energy involved during generating electricity from PV array. Then we can expect that there will be no abrupt changes on the frequency. However, load changes might lead to significant frequency changes that might affect the whole system. It is vital to manage this kind of problems to make sure that PV-based active generator in the micro-grid can operate efficiently (Table 14).

These are the things that need to be monitored and analysed. For a micro-grid-connected PV-based active generator, the network operators need reliable and robust PV energy output forecasting system in operational planning. PV array output depends not only on the incident solar radiation but also on the operating cell temperature as well as shading effect and its operating points. Therefore, a proper

Table 14 Timing classification for energy management system in Micro-grid (Kanchev et al. 2011)

Long term	Short term
• Electricity market	• Voltage control
• Load forecasting	• Frequency control
• Renewable energy production	• Dynamic storage availability
• Load management	• Power capability
• Energy storage availability	

forecasting methodology is required for predicting the PV array output. It is important to identify the pattern of the historical data set for predicting the output (Matallanas et al. 2012). There are a lot of work which have been done in this matter. This can be very helpful to generate enough power for PV-based active generator.

4.7 Conclusion

The utilization of PV as a source of electricity is something that needs to be emphasized now. The dependencies on the conventional way of generating energy via 'unclean' source and method need to be minimized. PV-based active generator can be a new way of generating energy in the future. It will be clean and very promising. Since this type of generator needs a good energy storage system, battery system with ultra-capacitor will be a great combination. Battery storage with the appearance of ultra-capacitor will increase the system efficiency as the battery will be able to store and release energy gradually, while ultra-capacitor effectively acts as storage device with very high power density. The new type of battery known as the VRB can be used in the near future for this type of generator.

For a better management in the micro-grid level, a new method of managing the energy needs to be implemented. The crucial part is to maintain the frequency and for this, a droop controller that will be connected with the micro-grid and the PV-based active generator needs to be developed. On the energy management side, the PV forecasting, power quality concern and fault issues need to be foreseen. Forecasting the PV energy needs to be done to make sure that there will be no shortage of power during operation, and if there is a power shortage, there should be a plan to overcome this problem.

In order to provide reliable energy, the micro-grid operator should monitor the power quality (harmonics, voltage variations). This is to avoid any deficiency to the consumer and this may lead to fault. Since in micro-grid environment there are probabilities on the islanding effect and this might as well affect the PV-based active generator, avoiding fault is something that the operator needs to be considered. Since the PV-based generator has a promising future, there should be more research on the energy storage side and on the PV cells. The impact of higher efficiency on PV cells and minimizing cost for a battery will have a significant impact to this new type of PV-based generator.

References

Aichhorn, A., Greenleaf, M., Li, H., & Zheng, J. (2012). *A cost effective battery sizing strategy based on a detailed battery lifetime model and an economic energy management strategy.* Paper presented at the Power and Energy Society General Meeting, 2012 IEEE.

Carvallo, A., & Cooper, J. (2011). *The advanced smart grid: Edge power driving sustainability*: Artech House.

Farhangi, H. (2010). The path of the smart grid. *Power and Energy Magazine, IEEE, 8*(1), 18–28.

Guasch, D., & Silvestre, S. (2003). Dynamic battery model for photovoltaic applications. *Progress in Photovoltaics: Research and Applications, 11*(3), 193–206.

Guerrero, J. M., Vasquez, J. C., Matas, J., de Vicuña, L. G., & Castilla, M. (2011). Hierarchical control of droop-controlled AC and DC microgrids—a general approach toward standardization. *IEEE Transactions on Industrial Electronics, 58*(1), 158–172.

Hill, C. A., Such, M. C., Chen, D., Gonzalez, J., & Grady, W. M. (2012). Battery energy storage for enabling integration of distributed solar power generation. *IEEE Transactions on Smart Grid, 3*(2), 850–857.

Kanchev, H., Lu, D., Colas, F., Lazarov, V., & Francois, B. (2011). Energy management and operational planning of a microgrid with a PV-based active generator for smart grid applications. *IEEE Transactions on Industrial Electronics, 58*(10), 4583–4592.

Kanchev, H., Lu, D., Francois, B., & Lazarov, V. (2010). *Smart monitoring of a microgrid including gas turbines and a dispatched PV-based active generator for energy management and emissions reduction.* Paper presented at the Innovative Smart Grid Technologies Conference Europe (ISGT Europe), 2010 IEEE PES.

Kunsman, S. A. Protective relaying and power quality.

Le Dinh, K., & Hayashi, Y. (2013). *Coordinated BESS control for improving voltage stability of a PV-supplied microgrid.* Paper presented at the Power Engineering Conference (UPEC), 2013 48th International Universities'.

Li, X., Hui, D., & Lai, X. (2013). Battery energy storage station (BESS)-based smoothing control of photovoltaic (PV) and wind power generation fluctuations.

Lu, D., & Francois, B. (2009). *Strategic framework of an energy management of a microgrid with a photovoltaic-based active generator.* Paper presented at the Advanced Electromechanical Motion Systems & Electric Drives Joint Symposium, 2009. ELECTROMOTION 2009. 8th International Symposium on.

Lu, D., Zhou, T., Fakham, H., & Francois, B. (2008). *Design of a power management system for an active PV station including various storage technologies.* Paper presented at the Power Electronics and Motion Control Conference, 2008. EPE-PEMC 2008.

Matallanas, E., Castillo-Cagigal, M., Gutiérrez, A., Monasterio-Huelin, F., Caamaño-Martín, E., Masa, D., & Jiménez-Leube, J. (2012). Neural network controller for Active Demand-Side Management with PV energy in the residential sector. *Applied Energy, 91*(1), 90–97.

Najy, W., Zeineldin, H., & Woon, W. (2013). Optimal protection coordination for microgrids with grid-connected and islanded capability.

Nguyen, T. A., Qiu, X., Gamage, T. T., Crow, M. L., McMillin, B. M., & Elmore, A. (2011). *Microgrid application with computer models and power management integrated using PSCAD/EMTDC.* Paper presented at the North American Power Symposium (NAPS), 2011

Quiggin, D., Cornell, S., Tierney, M., & Buswell, R. (2012). A simulation and optimisation study: Towards a decentralised microgrid, using real world fluctuation data. *Energy, 41*(1), 549–559.

Shah, R., & Mithulananthan, N. (2011). *A comparison of ultracapacitor, BESS and shunt capacitor on oscillation damping of power system with large-scale PV plants.* Paper presented at the 21st Australasian Universities Power Engineering Conference (AUPEC), 2011.

Teodorescu, R., & Liserre, M. (2011). *Grid converters for photovoltaic and wind power systems* (Vol. 29). New York: Wiley.

Vallvé, X., Graillot, A., Gual, S., & Colin, H. (2007). *Micro storage and demand side management in distributed PV grid-connected installations.* Paper presented at the EPQU 2007. 9th International Conference on Electrical Power Quality and Utilisation, 2007.

Wang, G., Ciobotaru, M., & Agelidis, V. G. (2012a). *Minimising output power fluctuation of large photovoltaic plant using vanadium redox battery storage.* Paper presented at the 6th IET International Conference on Power Electronics, Machines and Drives (PEMD 2012).

Wang, G., Ciobotaru, M., & Agelidis, V. G. (2012b). *PV power plant using hybrid energy storage system with improved efficiency.* Paper presented at the 2012 3rd IEEE International Symposium on Power Electronics for Distributed Generation Systems (PEDG).

Yoo, H.-J., Kim, H.-M., & Song, C. H. (2012). *A coordinated frequency control of Lead-acid BESS and Li-ion BESS during islanded microgrid operation.* Paper presented at the 2012 IEEE Vehicle Power and Propulsion Conference (VPPC).

Zhang, J., Wang, J., & Wu, X. (2012). *Research on supercapacitor charging efficiency of photovoltaic system.* Paper presented at the 2012 Asia-Pacific Power and Energy Engineering Conference (APPEEC).

PV Component Selection for Off-Grid Applications

Parimita Mohanty and Mukesh Gujar

Abstract Although component selection is one of the major tasks of the PV system design and installation, it is often overlooked and poorly integrated within the thought process. This particular aspect needs to be given special attention and awareness ought to be created amongst the installer. This chapter deals with the guidelines, methodology and approaches that need to be adopted for the appropriate selection of the components used in the solar PV-based off-grid application. These approaches are developed based on the state-of-the art system design as well as field experience and extensive research work carried out by the authors in this sector. This chapter will help readers in understanding the importance of each component of the solar PV system which may significantly affect its performance. Further, it will guide the implementers and designers in proper selection of the solar PV components for off-grid applications.

1 Introduction

Appropriate selection of components for a typical battery-based off-grid solar PV system is extremely important as they affect the system performance, efficiency, reliability, maintenance cost and aesthetics in the long run. Component selection and its overall integration are very critical for proper functioning and the reliability of the overall PV system. There would be a drastic reduction in the energy yield if the operating characteristics of different components do not match (Mohanty 2008). For example, if *voltage and current limits of the solar PV do not match with the*

P. Mohanty (✉)
The Climate Group, Nehru Place, New Delhi, India
e-mail: pmohanty@theclimategroup.org

M. Gujar
The Energy and Resources Institute, IHC Complex, Lodhi Road,
New Delhi 110003, India
e-mail: mgujar88@gmail.com

© Springer International Publishing Switzerland 2016
P. Mohanty et al. (eds.), *Solar Photovoltaic System Applications*,
Green Energy and Technology, DOI 10.1007/978-3-319-14663-8_4

inverter's voltage and current characteristics, the desired output will not be obtained.

Therefore, it is important to follow the guidelines for the selection of various components of a solar PV system for off-grid applications.

2 Guidelines for the Selection of Solar PV Components for Off-Grid Applications

A typical solar PV system for off-grid application consists of the following major components:

- PV Array;
- Charge controller;
- Inverter;
- Storage battery;
- PV source circuit combine box;
- PV fuse disconnect;
- Ground Fault protection circuit;
- Battery fuse disconnect;
- Inverter fuse disconnect and
- Cables, wires and other accessories.

The following sections discuss the critical points to be considered for selecting the above components of a PV system.

3 Guidelines for the Selection of PV Modules

Solar PV modules are subjected to various national and international standards; however, these only provide measures of safety of the devices and do not indicate field performance or reliability. Therefore, the following parameters should be considered for module procurement to ensure their reliable field level performance:

- Electrical characteristics and its specification;
- Temperature tolerance and hail impact resistance;
- Efficiency, dimensions and weight;
- System compatibility;
- Quality requirement, quality marks, standards and specifications and
- Warranties and guarantees.

Each of the above -mentioned parameters is explained below.

3.1 PV Module Electrical Characteristics

While buying a solar module, one should ask for the I–V characteristic curve (Current versus Voltage Graph) of the solar module. When buying multiple solar modules, I–V characteristic curve for each should be taken from the supplier. An I–V curve can be used to ascertain the following:

- Open-Circuit Voltage (V_{oc}) and Short-Circuit current (I_{sc});
- Maximum Power Point (MPP);
- Wattage at Maximum Power Point (W_p);
- Voltage at Maximum Power Point (V_m);
- Current at Maximum Power Point (I_m) and
- Fill Factor (FF).

For a solar power plant, modules with higher wattage, i.e. higher W_p, should be used because it will require less number of modules thus covering less roof area, fewer mounting structure and fewer inter-module electrical connections (Fig. 1).

The Fill Factor (FF) of the solar module is a critical parameter, although many people tend to ignore it. The FF is defined as the ratio of the maximum power (W_p) from the PV module to the product of the open-circuit voltage (V_{oc}) and short-circuit current (Isc). Graphically, FF is the measure of the squareness of the I–V characteristic curve of a PV module, and is given by the area of the largest rectangle which will fit in the I–V curve. FF ranges from 0.76 to 0.8, in case of a good solar PV module (Fig. 2a). So by knowing the FF of a solar PV module, the quality of a PV module can be found out [good quality (Fig. 2a) or bad quality (Fig. 2b)].

Power tolerances are also very critical and these need to be checked before the purchase of the module as the module's output varies significantly based on the tolerance range. Generally, power tolerance of a good quality PV module is ±3 % to ±5 %.

Fig. 1 Electrical characteristic of the solar PV module

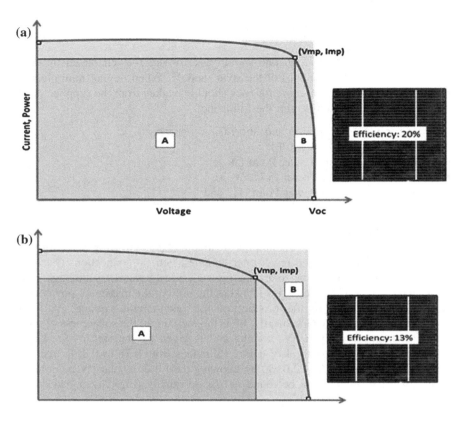

Fig. 2 **a** Fill factor of a good quality solar PV module with higher efficiency, **b** fill factor of a bad quality solar PV module with lower efficiency

3.2 PV Module Temperature Tolerance and Hail Impact Resistance

The open-circuit voltage (V_{oc}) of an individual silicon solar PV cell reduces by 2.3 mV per degree rise of cell temperature. Therefore, the temperature coefficient of the voltage is negative and expressed as shown in Eq. 1:

$$\frac{dV_{OC}}{dT} = -2.3\text{mV per }°\text{C for Si solar PV cell} \tag{1}$$

As a PV module consists of large number of PV cells in series, the voltage drop increases proportionally as is shown in Eq. 2:

$$\frac{dV_{OC}}{dT} = -2.3\text{mV per }°\text{C for Si solar PV module} \tag{2}$$

where n_c is the number of PV cells in series.

On the other hand, the temperature coefficient of the current is slightly higher and the current increases very slightly with an increase in the cell temperature (Eq. 3). However, this is a small effect and the temperature dependence of the short-circuit current from a silicon solar cell is

$$\frac{1}{I_{SC}}\frac{dI_{SC}}{dT} \approx 0.0006 \text{ per }°\text{C for Si} \tag{3}$$

However, the temperature coefficient of the power is negative for the silicon solar PV module. This means, with the rise of temperature, that the maximum power output of the silicon solar PV module decreases which can be calculated as per the following equation (Eq. 4):

$$\frac{1}{P_{Max}}\frac{dP_{Max}}{dT} \approx -(0.004 \text{ to } 0.005) \text{ per }°\text{C for Si} \tag{4}$$

This temperature coefficient effect needs to be taken care of in the system design and in estimating the yield of a PV array. If the power output of the silicon PV module is specified as 100 Wp at Standard Test Condition (STC), in real condition, its actual power output at 50 °C would be 90 Wp.

3.3 PV Module Efficiency, Dimensions and Weight

PV module efficiency has a practical relevance because a module with a low efficiency will simply cover a larger roof space to provide the same power output compared to an efficient module. So if the cost per Watt peak (Wp) of the solar PV module is substantially lower for the relatively lower efficiency module and there is a considerable free surface area already available without any future plan for utilizing that roof space, then it makes sense to install relatively inefficient modules. However, if the low-cost modules lead to reduced life and poor performance, then it is better to go for high efficiency PV modules even at a higher price. Therefore, modules with same Wp but different efficiencies can be compared with their cost and the available surface area. The most optimum PV module can thus be chosen.

3.4 PV Module with System Compatibility

PV module or array should be compatible with the rest of the electrical components in the system. Particularly, the PV module or array's electrical characteristics should be compatible with that of the inverter, charge controller and the batteries. For example, the voltage of the PV array should match with the battery voltage in order to be able to charge the battery appropriately and/or it should match with the input voltage required for the inverter in order to provide the desired AC output.

3.5 PV Module Quality Requirement, Quality Marks, Standards and Specifications

PV modules which comply with appropriate national and international standards developed by recognized institutions such as the International Electro-technical Commission (IEC), the American Society for Testing and Material (ASTM) and the Sandia National Laboratory can be considered as reliable and are likely to have longer life (Antony et al. 2007). The compliance with the standards will also take care of the minimum degradation of power output (%) from the solar PV module in a defined time period.

Similarly, the ISO 9001:2000 standard is the globally accepted system for quality management of the manufacturing processes as well as for distributors, system assemblers and installers. This is a generic standard, used by many industries. Several PV module manufacturers presently meet the ISO 9001:2000 standard and are certified. Further, the Global Approval Program for Photovoltaics (PV GAP) was established in 1998 under the auspices of an independent organization, PV GAP, to promote the use of international standards, quality management processes and organizational training in the manufacture, installation and sale of PV systems. It also defines the procedure for certification of photovoltaic products and the use of the PV quality mark and PV quality seal.

3.6 PV Module Warranties

PV module warranties are extremely critical and the modules with a minimum guaranteed output of 80 % of its original rating after 20–25 years of installation should be purchased.

3.7 Performance Comparison of Different PV Technologies

Table 1 compares the performance of different PV technologies which can be used for designing the various PV systems as well as for selecting the appropriate PV modules for specific site conditions and requirements.

4 Guidelines for Battery Selection

Solar PV modules can generate power only when it is exposed to the sunlight. Therefore, there is a need for storage batteries to store the energy that is being generated by the PV module during periods of high irradiance and make it available at night as well as during overcast periods. Unlike car batteries, solar PV applications demand for frequent charging and discharging of batteries. So the type of battery used here is not the same as that of an automobile battery. Figure 3 shows the different categories of batteries.

Out of the above-mentioned batteries, the following types of batteries are available and mostly used for PV application:

- Lead acid battery (most common);
- NiCd (Nickel Cadmium);
- NiMH (Nickel Magnesium Hydroxide) and
- Lithium-based battery.

A large numbers of batteries are available in the market and each of these is designed for a specific application. Lead acid batteries are most widely used in solar PV applications. Currently, lithium-ion based batteries are also being used in solar PV applications, particularly for small-scale applications. Other battery types, such as NiCd or NiMH, are used in portable devices. The life of typical solar system batteries spans from 3 to 5 years. However, a life of around 6–8 years can be achieved with proper battery management and regular preventive maintenance of the battery bank.

While selecting the batteries, utmost attention ought to be given to ensure that all the batteries used in a battery bank must be of same type, same manufacturer, same age and must be maintained at equal temperature. Further, the batteries should have the same charge and discharge properties under these circumstances. If the above characteristics do not match, there is a high probability of huge energy loss within the battery bank.

For product designers, an understanding of the factors affecting the capacity and life of the battery is very important in order to manage the product performance and its warranty. Following characteristics need to be ensured while selecting a battery used for solar applications:

- Battery capacity and discharge rate;
- Cycle life and temperature;

Table 1 Performance of different PV technologies

	Mono-crystalline	Poly- crystalline	Thin-film		
			Amorphous	Cadmium Telluride (CdTe)	Copper Indium Gallium di-selenide (CIGS
Typical module efficiency at the field	13–18 %	10–14 %	6–8 %	9–11 %	10–12 %
Lab scale Solar PV cell efficiency	25.0 %	20.4 %	13.4 %	18.7 %	20.4 %
Space required for installation of 1 kWp of solar PV array (in m²)	6–9	8–9	13–20	11–13	9–11
Typical length of warranty	25 years	25 years	10–25 years		
Temperature resistance	Performance drops by 10–15 % from its value at standard test condition at high temperatures	Less temperature resistant than mono-crystalline	Tolerates extreme heat	Relatively low impact on performance	
Additional details	Oldest cell technology and most widely used	Less silicon waste in the production process	Tend to degrade faster than crystalline-based solar PV modules		
I–V Curve Fill Factor (Idealized PV cell is 100 %)	73–82 %		60–68 %		
Module construction	With Anodized Aluminium		Frameless, sandwiched between glass; Lower cost, lower weight		
Inverter Compatibility and Sizing	Lower temperature coefficient is beneficial		System designer has to consider factor such as temperature coefficients and $V_{oc} - V_{mp}$ difference		
Mounting systems	Industry standard		Special clips and structures may be needed		

Source Authors compilation, 2014

- Battery Ampere-hour (Ah) efficiency;
- High charging current capability;
- Good reliability under cyclic, deep discharge conditions

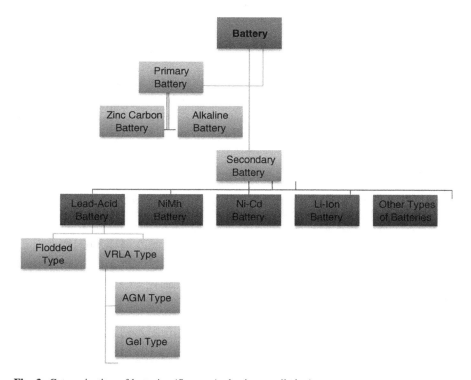

Fig. 3 Categorization of batteries (*Source* Author's compilation)

- Good power density, high recharge efficiency, rapid re-chargeability;
- Maintenance free, wide operating temperature range; and
- Low cost per Ah, low self-discharge rate.

4.1 Battery Capacity

The capacity of a battery is measured in Ampere-hour (Ah). It provides information on the number of hours a specific current can be delivered by a fully charged battery before it gets completely discharged.

Battery capacity depends on the following:

Discharging current: the higher the discharging current, the lower the capacity, and vice versa. For example, a battery delivering 1 Ampere current for 100 h has a capacity of 100 Ah. However, if the same battery is discharged (delivering a current) with 8 Amperes current, it may provide that current for 10 h. This means that the capacity is no more 100 Ah and instead reduces to 80Ah. *Therefore, from the designer's perspective, the selection of the battery based on its capacity is*

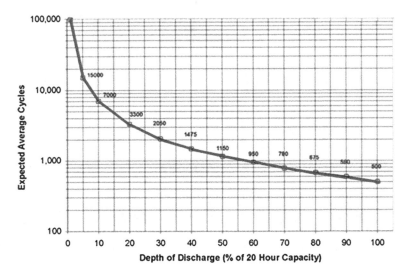

Fig. 4 Depth of discharge versus number of cycles of battery discharge (*Source* author's compilation)

important but more importantly at what rate (C/10 or C/20 or C/100) is most critical (Lasnier and Ang 1990). *Because that decides what would be the actual capacity of the battery if it is not discharged at the designed rate.*

To supply a given load of X Watt, a bigger battery will be drained by a lower percentage, and would last for more cycles. This means that, to supply a 50 Ah load, a 100 Ah battery would be drained by 50 %, whereas a 200 Ah battery would only be drained by 25 %. The bigger battery will last longer. From Fig. 4, it can be seen that if the depth of discharge (DoD) changes from 50 to 25 %, then the life cycle of the battery would be more than doubled. Thus a bigger battery is always recommended. But a bigger battery would also increase the cost. Thus a balance between the cost of the battery and its capacity can be made to select the most appropriate capacity and type of battery.

Temperature: The capacity of the battery available in the market is specified from 25 to 27 °C. When the temperature exceeds this range, the capacity of the battery increases with temperature, whereas the life of the battery decreases. The performance of all batteries drops drastically at low temperatures. At minus 20 °C (−4 °F), most nickel-, lead- and lithium-based batteries stop functioning. Specially built lithium ion brings the operating temperature down to minus 40 °C, but only on a reduced discharge. So for lower temperature, lithium-ion battery can be chosen.

Figure 5 shows the ideal working temperature of the battery in order to get the optimum life cycle. Operating above 55 °C damages the battery permanently. One of the main functions of the battery management system is to keep the battery cells operating temperature within their designed operating temperature window.

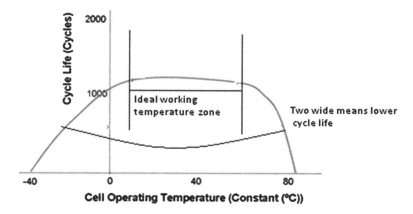

Fig. 5 Operating temperature of the battery versus cycle life

4.2 Life of a Battery

The life of lead acid batteries is quoted as 5 years, 10 years or 20 years. However, this is a *conditional statement,* and generally this condition is overlooked. The condition is that the life of the battery will be X years if the battery is *"kept in a specific temperature band"* and is *"kept at a specific float voltage".* If these conditions are not satisfied, the life of the battery will reduce and the rate of reduction of life depends upon how far its operation deviates from the designed values. Therefore, it is always advisable to operate the batteries near to its specified conditions, particularly in terms of its temperature band and float voltage.

4.3 Self-discharge of a Battery

This parameter is typically critical for a remote electrification project, where the transportation of the materials including batteries takes more time due to the remote and inaccessible nature of the site. If the self-discharge rate of the battery is high, it needs regular charging, even in unused and no load conditions, in order to extract the desired energy output and life of the battery. Therefore, based on the type of batteries selected and their self-discharge rate, the frequency of charging the battery in no load condition can be decided. The typical self-discharge rates for common rechargeable batteries are as follows:

- Lead Acid batteries—4 to 6 % per month;
- Nickel Cadmium batteries—15 to 20 % per month;
- Nickel Metal Hydride batteries—30 % per month and
- Lithium-Ion batteries—2 to 3 % per month,

It shows that if a new fully charged, unused lead acid battery gets fully discharged in 15–20 days, the new fully charged, unused Nickel Metal Hydride battery will get fully discharged in only 3–4 days.

4.4 Comparison of Batteries Available in the Market

Table 2 compares the characteristics of four commonly used rechargeable batteries used for solar applications.

Based on the above-mentioned parameters, the appropriate battery for a particular application can be selected. For example, if the battery is to be used for a location which is remote with very difficult terrain and road connectivity, then in such a case, lithium-ion battery can be used because of its quite high cycle life as compared to other types of batteries and thus requiring less frequent replacements. At the same time, it requires very less maintenance. Further, unlike a lead acid battery, a lithium-ion battery can be discharged up to 98–99 %, without shortening the life of the battery and thus reducing the size of the battery by half compared to that of a lead acid battery for a specific application. However, the charge controller

Table 2 Comparison of characteristics of different batteries

Specification	Lead Acid	NiCd	NiMH	Lithium based		
				Cobalt	Manganese	Phosphate
Specific energy density (Wh/kg)	30–50	45–80	60–80	150–190	100–135	90–120
Cycle life (number of cycles)	200–300	1000	300–500	500–1000	500–1000	1000–2000
Duration of charging	8–16 h	1 h	2–4 h	2–4 h	1 h or less	1 h or less
Overcharge tolerance	High	Moderate	Low	Cannot tolerate		
Self-discharge rate per month	5 %	20 %	30 %	<5 %		
Frequency of maintenance requirement	In every 3–6 months (topping)	In every 30–60 days – recharging	In every 60–90 days - recharging	Normally not required		
Cell voltage	2 V	1.2 V	1.2 V	3.6 V	3.6 V	3.3 V
Temperature range during charging	–20 to 50 °C	0–45 °C				
Safety requirement	Thermally stable	Thermal protection mandatory				

of such a battery needs to be designed properly with proper temperature control and thermal protection circuits as well as overcharge protection as the safety measure.

On the other hand, if the battery is to be used for very large rugged applications under varying operating temperatures and without proper sophisticated electronic charge controllers, a lead acid battery can be preferred over other batteries.

5 Guidelines for the Selection of Inverters

The following sections describe the selection of inverters based on different characteristics.

5.1 Selection of Inverters Based on Its Configurations

5.1.1 Single-Stage/Central Inverter

The single-stage inverter (central inverter) is widely used for large-scale power applications. Here, the single power processing stage takes care of all the tasks of Maximum Power Point Tracking (MPPT), voltage amplification and grid side current control. In this configuration, the solar modules are connected in series to create strings with output voltage high enough to avoid an additional voltage boost stage (Tang and Zhao 2010). In order to obtain the desired power level, the strings are connected in parallel through interconnection diodes (string diodes) as shown in Fig. 6.

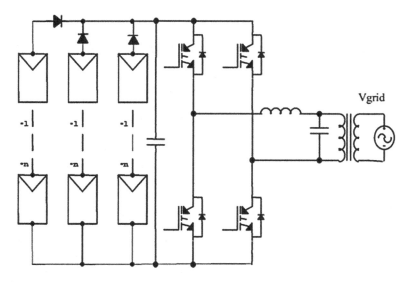

Fig. 6 Connections of solar PV strings

5.1.2 Double- or Multistage Inverter

Here, each string is connected to a double- or a single-stage inverter. If a large number of modules are connected in series to obtain an open-circuit voltage higher than 360 V, the DC/DC converter can be eliminated. On the other hand, if a few number of PV modules are connected in series; a DC/DC boost converter is used. The DC–DC converter is responsible for the MPPT and the DC–AC inverter controls the grid current.

5.1.3 Multi-string Multistage Inverters with High-Frequency Transformer

Another topology adopted is multi-string, multistage inverter. The multi-string inverter has been developed to combine the advantage of higher energy yield of a string inverter with the lower costs of a central inverter. Lower power DC/DC converters are connected to individual PV strings. Each PV string has its own MPPT, which independently optimizes the energy output from each PV string. All DC/DC converters are connected via a DC bus through a central inverter to the grid. Depending on the size of the string the input voltage ranges between 125 and 750 V, here, the system efficiency is higher due to the application of MPPT control on each string and higher flexibility comes from the ease of extensions for the photovoltaic field. This topology is more convenient for power levels below 10 kW. The multi-string inverters provide a very wide input voltage range (due to the additional DC/DC stage) which gives the user better freedom in the design of the PV system. However, the disadvantages are that it requires two power conversion stages to allow individual tracking of the inputs.

5.2 Selection of Inverters Based on Switching Devices

To effectively perform Pulse Width Modulation (PWM) control for the inverter, Insulated Gate Bipolar Transistor (IGBT) and Metal Oxide Semiconductor Field Effect Transistor (MOSFET) are mainly used as switching devices. *IGBT is used in more than 70 % of the surveyed inverter products and MOSFET is used in around remaining 30 % of the inverter.* As far as differences in characteristics between IGBT and MOSFET are concerned, the switching frequency of IGBT is around 20 kHz and it can be used for large power capacity inverters exceeding 100 kW (Kjaer et al. 2005). On the other hand, although the switching frequency of MOSFET can go up to 800 kHz, its power capacity is reduced at higher frequencies and MOSFET are used for output power range between 1 and 10 kW. So in a nutshell, both IGBT and MOSFETs are used for small to medium range PV system with power capacity of 1–10 kW, whereas IGBTs are used for large-scale power plants with power rating of 100 kW or more. High-frequency switching can reduce the harmonics in the output current, the size and the weight of an inverter and thus nowadays High-Frequency (HF) inverters with a compact size are available and widely used.

5.3 Selection of Inverters Based on Operational Perspectives

In order to assess the inverter's performance in terms of its operational perspective, a literature review and collection of secondary information were carried out by the authors. For this analysis, information of about 200 models of different inverters with different capacities and types were collected. The subsequent sections (Sect. 4.5.5–4.5.13) bring the findings from that survey.

5.4 Features of Grid Connectivity

The distributed or off-grid inverter should have the grid connectivity feature (both incoming and outgoing) so that these solar PV systems would not be completely obsolete when the grid extension takes place. As there are massive plans for conventional rural electrification, it is always wise to select an inverter having grid connectivity features from the beginning with some incremental cost. It is cheaper than completely changing the inverters later on.

5.5 AC Voltage and Frequency Range

An inverter can be operated without any problems within the tolerance of +10 and − 15 % of the standard voltage, and ±0.4 to 1 % of the normal frequency specified by the grid standards of any country. For example, in India, where the single-phase AC line is specified as 230 V, 50 Hz, the inverter should work at any voltage value between 253 and 198 V and at any frequency value between 49.5 and 50.5 Hz without any problem. Any inverter which does not have this wide range might not be considered, particularly for distributed power systems which are installed in relatively remote and rural locations, where wide fluctuations of voltage and frequency are prevalent (Kim et al. 2009)

5.6 Operational DC Voltage Range

The operable range of the DC voltage differs according to the rated power of the inverter, the rated voltage of the AC utility grid system and the design policy. In this survey, the operable range of the DC voltage for a capacity in the range of

180–500 W includes 14–35 V, 30–60 V. Similarly, the operable DC voltage range for a capacity of 10 kW or over includes 330–1000 V. Hence, depending on the operational range of the voltage range of the inverter, the capacity and configuration of the solar PV modules should be decided. While designing the capacity of the solar PV system, this is one criterion that decides how many modules need to be connected in series or parallel to get the required DC operating voltage.

5.7 AC Harmonic Current from the Inverter

Minimization of harmonic current production is required as the harmonic current adversely affects load appliances connected to the distribution system and can impair load appliances when the harmonic current is increased (Farahat et al. 2012). The results of this survey show that the Total Harmonic Distortion (THD)[1] is 3–5 %. However, there are certain inverters in the power rating of 10–100 kW that have THD in the range of 1–5 %.

5.8 Inverter Conversion Efficiency

Figure 7 presents the performance of a range of inverters.

Figure 7 shows that the euro efficiency range for all the inverters varies from 94.5 to 98.7 % (Photon International 2012). In the medium-scale range (10–20 kW), there are several inverters available with the euro efficiency range of 97–98 % and the incremental cost of these inverters is not much different from that of the low efficiency inverters (generally USD 50/kW). Thus the project designer can evaluate the cost versus benefit of the inverter in terms of the enhanced efficiency.

5.9 Operational Environment

The installation conditions of the inverter (the indoor installation specification or the outdoor installation specification), the ambient temperature, the requirements for water and dust proofing, actual audible noise level of the inverter and applicable regulations for EMC (electro-magnetic compatibility), need to be examined carefully. As per the survey (Fig. 8), the maximum acceptable ambient temperature at nominal AC power is in the range of 40–75 °C. Whereas this range is relatively wider for 10–40 kW inverter, it is narrower in the operational range of 40–50 °C for larger inverters.

---

[1] the total distortion factor of the current normalized by the rated fundamental current of many of the inverters.

Fig. 7 Power rating versus euro efficiency of inverters

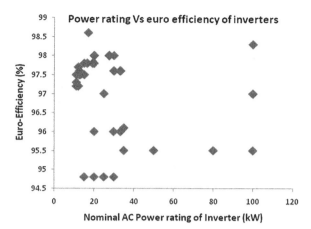

Fig. 8 Maximum acceptable temperature at nominal AC power

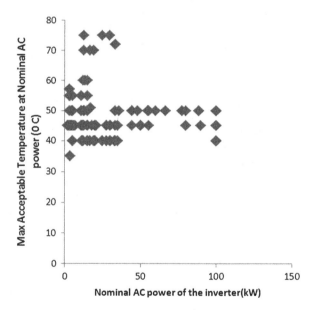

5.10   Required Protection Devices or Functions

Protective functions include protection for the DC and the AC sides. The protective functions for the DC side include those for DC over power, DC over voltage, DC under voltage, DC over current and detection of DC grounding faults. Protective functions for the AC side include AC over voltage, AC under voltage, AC over current, frequency increase, frequency drop and detection of AC grounding. Most of the inverters include these basic protections.

5.11 Standby Power Consumption

The standby power consumption of the inverter is a very important parameter that needs to be checked, particularly for off-grid PV applications where the only source for providing power is the PV. So, by practice the lesser the standby power consumption is, the better it is for distributed PV applications. As per Fig. 9, it is observed that there is a wide range of products that are available with normalized standby power consumption for inverter capacity of less than 20 kW. So the inverter can be judiciously selected so that self-consumption of these devices would not be significant.

5.12 Inverter System Cost, Size and Weight

The cost of the inverter system is very crucial when considering the economy of a photovoltaic power system. Although the cost of the inverter varies from country to country as well as with the make and model, the average cost range of the inverter is found to be USD 600–1000 per kW (Photon International 2012). The weight of the inverter system differs considerably according to the presence/absence of the isolating transformer (shown in Fig. 10). However, from the project developer's point of view, it is a very critical parameter to judge as the weight of the system affects the total transportation cost, handling and installation cost. For an inverter for a household PV power system, the weight reduction is important when the inverter is installed indoors or it is to be mounted and thus an appropriate inverter with lower weight should be preferred.

Fig. 9 Normalized standby power consumption of a selection of inverters

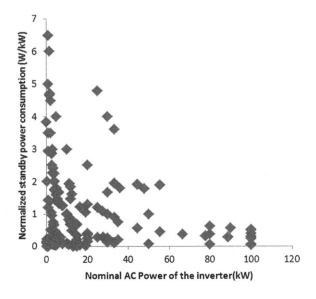

Fig. 10 Normalized weight
of different inverters

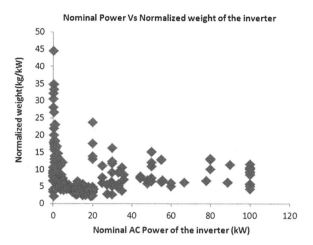

5.13 System Guarantee

System guarantee plays a crucial role as it influences the entire system economics. Although the cost data for each of the inverters is not available, it is noted that there are some inverters that are available with an extended guarantee of 25 years.

6 Selection of Protective Devices

Besides the major component selection, safety measures are equally important for the solar PV system for which the protective devices to be used need to be selected properly. Failure to do so may not only lead to system efficiency but also may pose severe safety hazards. Typical protection devices which are to be selected carefully are as follows:

1. PV source circuit combine box and PV fuse disconnect;
2. Battery fuse disconnect;
3. Inverter fuse disconnect;
4. Ground Fault protection circuit;
5. Lighting arrestor;
6. Grounding/Earthing;
7. Surge protector and
8. Cables, wires and other accessories.

Each of the components is described briefly in subsequent sections.

6.1 PV Source Circuit Combine Box and PV Fuse Disconnect

These fuses or circuit breakers [both known as over current protective devices (OCPD)] are installed to protect the PV modules *and* wiring from excessive reverse current flow that can damage PV cell interconnects and the wiring between the individual PV modules.

The rating of the fuse is specified by the PV module manufacturer. As per the NEC requirement for over current protection, the fuse rating marked on the back of the PV module must be at least 156 % of short-circuit current (Isc) of the PV module at STC. The fuse will generally be a *dc-rated cartridge-type fuse* that is installed in a finger-safe pull out-type fuse holder.

6.2 Battery Fuse Disconnect

The battery disconnect is a switch or a circuit breaker used for overcurrent protection and must be able to interrupt any battery short-circuit current. The dc voltage ratings of all components should be based upon the maximum system voltage, which is the PV system open-circuit voltage multiplied by the appropriate correction factor from NEC 690.7.

Since the battery is a dc circuit, dc-rated components should be used. It is important to remember that dc circuits require dc-rated components. It is not acceptable to substitute ac fuses, disconnects or circuit breakers for dc applications unless and until these are rated for both ac and dc circuits.

6.3 Grounding

Grounding is one of the most critical tasks in the entire installation of a solar PV system. Grounding means connecting a part of the system's structure and/or wiring electrically to the earth. Grounding of a system does four things:

- It discharges accumulated charges so that lightning is not highly accumulated in the system.
- If lightning does strike, or if a high charge does build up, the ground connection provides a safe path for discharge directly to the earth rather than through the wiring.
- It reduces shock hazard from the higher voltage (AC) parts of the system and
- It reduces electrical hum and radio caused by inverters, motors, fluorescent lights and other devices.

Grounding is required by the national electrical code (NEC). If the maximum system voltage of a PV system is greater than 50 V, then one conductor must normally be grounded. As per *NEC* 690.43, all exposed non-current-carrying metal parts of the components are to be grounded in accordance with *NEC* 250.134 or 250.136(A), regardless of the system voltage. This means that even if the current-carrying conductors of the PV array do not need to be grounded, all non-current-carrying metal equipment parts and cases must still be grounded with equipment-grounding conductors.

Where the ground is moist (electrically conductive), grounding can be provided by a copper-plated rod, usually 8 ft long, which is dug inside the earth. Where the ground is dry, especially sandy or where lightning may be particularly severe, more rods can be tied together through bare copper wire and installed.

If the PV array is at a distance from the house, ground rods can be dug near the house and a bare wire can be buried in the trench with the power lines.

6.4 Lightning Arrestor

In locations susceptible to lightning strikes, a lightning protection system must be provided, and all the exposed metallic structures of the solar PV system must be bound to the earthing system, and structures and PV module frames must be properly grounded. In certain geographical locations, solar PV systems might be exposed to the threat of lightning strikes. As lightning can cause damage to the PV modules and inverters, extra care must be required to ensure that proper lightning protection is provided for the solar PV system and the entire structure. The inverters should be protected by appropriately rated surge arrestors on the DC as well as AC side.

6.5 Surge Protector

Surge protection devices bypass the high voltages induced by lightning. They are recommended for additional protection in lightning-prone areas or where good grounding is not feasible (such as on a dry rocky mountain top), especially if long lines are being run to an array, pump, antenna, or between buildings. To reduce the possibility of a fire and to protect the system from a damage caused by lightings, it is recommended to have a voltage-clamping device across the DC bus bar. A metal oxide varistor (MOV) is commonly used in such applications.

6.6 Cables and Wires

PV array wiring should be done with minimum lengths of wire and tied into the metal framework and then run through a metal conduit. A rule of thumb is to limit the voltage drop from the array to power inverter to 2.5 % or less.

Positive and negative wires should be run together wherever possible, rather than being kept some distance apart. This will minimize induction of lightning surges.

7 Conclusion

Appropriate selection of each component of solar PV system is equally important and admittedly one of the most significant parts in implementation of such system for off-grid applications. The chapter has assisted the readers in understanding the importance of each component of the solar PV system, i.e. solar PV module to battery to inverter and to cable and wiring to other protection devices, which may significantly affect the overall performance. The chapter has also demonstrated the steps to be followed for proper selection of various components which would guide the implementers and designers to take a well-informed decision while designing and formulating the solar PV-based off-grid projects.

References

Antony, F., Durschner C., & Remmers, K.-H. (2007). *Photovoltaic for professionals—solar electric systems marketing, design and installation*. Solarpraxis AG publishers in association with Earthscan, London.

Kim, H., Kim, J., Min, B., Yoo, D., & Kim, H. (2009). A highly efficient PV system using a series connection of DC–DC converter output with a photovoltaic panel. *Renewable Energy, 34*(11), 2432–2436.

Kjaer, S., Pedersen, J., & BlaaBjerg, F. (2005, October). A review of single phase grid connected inverters for photovoltaic modules. *IEEE Transaction on Industry Applications, 41*(5), 1292–1306.

Lasnier, F., & Ang, T. G. (1990). *Photovoltaic engineering handbook*. Bristol: IOP publishing.

Farahat M. A., Metwally H. M. B., Ahmed A. B. D., & Mohamed, E. (2012). Optimal choice and design of different topologies of DC-DC converter used in PV systems, at different climatic conditions in Egypt. *Renewable Energy, 43*, 393–402.

Mohanty, P. (2008). Solar Photovoltaic technology-Renewable energy Engineering and Technology. New Delhi: TERI publication.

Photon International, July and August issue (2012).

Tang, Y., & Zhao, L. (2010, June). Maximum power point tracking techniques for photovoltaic array. *New Energy* 48–51.

Performance of Solar PV Systems

Tariq Muneer and Yash Kotak

Abstract This chapter provides a review of the complete solar photovoltaic system that includes a discussion on the trends in manufacturing technology of the constituting components. The chapter then provides the reader with a detailed discussion on the steps that a designer will need to assess the performance of the PV system. The chapter finally presents four case studies of real installations and indicates lessons learned from them. Throughout those case studies material is provided on detailed measurements related to the incident solar radiation, PV cell temperature, thermal losses that occur from PV modules and their measured efficiency. Thus, it is possible to evaluate the accuracy of the modelling procedures using the latter measured data set.

1 Introduction

This chapter provides a review of entire PV system and its components. Furthermore, detailed designing and modelling procedures for the PV systems are presented. The chapter then enumerates lessons learned from four case studies of PV systems that range from a few kilowatts to hundreds of kilowatts.

T. Muneer (✉)
Edinburgh Napier University, 10 Colinton Road, Edinburgh EH10 5DT, UK
e-mail: T.Muneer@Napier.ac.uk

Y. Kotak
Heriot Watt University, Riccarton, Edinburgh EH14 4AS, UK
e-mail: yk78@hw.ac.uk

© Springer International Publishing Switzerland 2016
P. Mohanty et al. (eds.), *Solar Photovoltaic System Applications*,
Green Energy and Technology, DOI 10.1007/978-3-319-14663-8_5

107

2 PV System Modelling and Design

This study presents a generalized PV system simulation which incorporates solar irradiance and weather data to the final energy outcome as power.

2.1 PV Output Modelling

In order to estimate the performance of PV systems radiation data are required. These are available on a monthly averaged basis at National Aeronautics and Space Administration (NASA) renewable energy resource website (Surface meteorology and solar energy) (NASA 2014) and temperature data from TuTiempo website (TuTiempo 2014).

Solar radiation incident on any given surface can be decomposed into two components, the direct or beam component emanating from the sun, and a diffuse component that results from multiple reflections and scattering due to particles in the atmosphere and reflection. The diffuse component may also include reflections from the ground and local surroundings, where the surface in question is sloped rather than horizontal. Differentiating between the two components is vital for accurate calculations in most solar energy applications; however, a number of steps may be required to arrive at realistic estimates at an appropriate level of detail for a given location depending on the basic data that is available.

For this example, the monthly average global radiation on the horizontal plane was made available for Edinburgh by the NASA website. Figure 1 shows the computational flow for any general surface, that is, one, which may have a given orientation and slope. A detailed flow chart of the model is shown in Fig. 2.

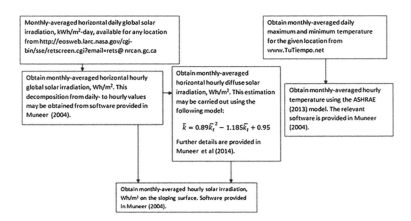

Fig. 1 Steps to generate slope solar irradiation. *Note* $\bar{k}_t$ = monthly average global clearness index, $\bar{k}$ = diffuse solar ratio. Refer to Muneer (2004) and Muneer et al. (2014)

Fig. 2 Flow chart for obtaining PV power output. *Note* Vm = voltage at maximum power point, Im = current at maximum power point, Isc = short circuit current, Voc = open circuit voltage

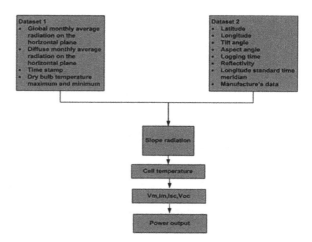

2.2 Factors Affecting PV System Performance

2.2.1 Irradiance

Irradiance has a direct effect on PV module output. An increase in irradiance increases the PV output. Irradiance also affects the PV cell temperature and thus affects the PV output.

The light-generated current is proportional to the flux of photons. Thus, an increase in irradiance proportionately increases the current. The voltage varies only logarithmically though with irradiance and its variation is therefore neglected in practical applications. Figures 3 and 4 show the influence of irradiance on the Current-Voltage (IV) characteristics and power output.

Fig. 3 Influence of irradiance on the IV characteristics

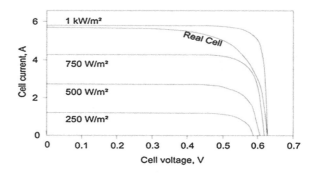

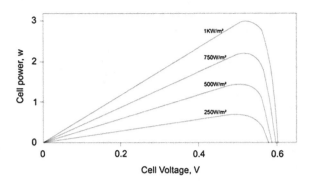

Fig. 4 Influence of irradiance on power output

2.2.2 Cell Temperature

Cell Temperature has an important effect on the power output from the cell as shown in Figs. 5 and 6.

The most significant is the temperature dependence of the voltage which decreases with increasing temperature. The voltage decrease of a silicon cell is typically 2.3 mV per degree C temperature rise. The temperature variation of the current is less pronounced.

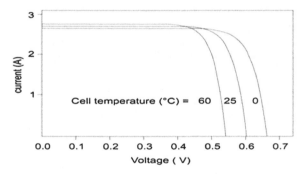

Fig. 5 Effect of cell temperature on the power output (Schematic A)

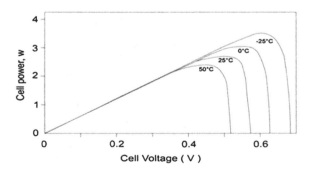

Fig. 6 Effect of cell temperature on the power output (Schematic B)

When PV cells are mounted on a module, they can be characterized as having a nominal operating cell temperature (NOCT) which is defined for open circuit conditions when the ambient temperature is 20 °C at AM 1.5 irradiance conditions, with an irradiance of 800 W/m^2 and wind speed of 1 m/s (Messenger and Ventre 2004).

The combined effect of irradiance and ambient temperature variation should be considered. A typical drop in the open circuit voltage of a silicon cell is 2.3 mV/C and hence the latter drop in a module with n cells will be 2.3 n mV/C. Thus, for example, in a typical 36-cell module which has an NOCT of 40 °C with an open circuit voltage, Voc = 19.4 V, Tc will rise to 55 °C when the ambient temperature rises to 30 °C and irradiance increases to 1000 W/m^2. This will produce a drop in Voc to 18.16 V, a 6 % decrease (Markvart 2000).

2.2.3 Solar Altitude and Solar Spectrum

Solar altitude and solar spectrum influence the solar irradiance intensity thus, γ_s.

As the sun moves through the sky, the elevation angle changes during the day and over the course of the year. When the solar altitude is perpendicular to the Earth, the sunlight takes the shortest path through the Earth's atmosphere. However, if the sun is at a shallower angle, the path through the atmosphere is longer. This results in greater absorption and scattering of solar radiation and, hence, lowers radiation intensity. The air mass factor (AM) specifies how many times the perpendicular thickness of the atmosphere the sunlight has to travel through the Earth's atmosphere. A simple relationship between solar altitude (height) γ_s and air mass is,

$$AM = \frac{1}{\sin \gamma_s} \qquad (1)$$

When the solar altitude is perpendicular ($\gamma_s = 90°$), AM = 1. This corresponds to the solar altitude at the equator at noon during the spring or autumn equinox.

The photoelectric conversion efficiency depends on matching of the incident spectrum with the cell's spectral response and the actual operating cell temperature. The change in power output as a function of the spectrum (AM) for PV modules based on crystalline and amorphous silicon is shown in Fig. 7.

3 Actual Performance Evaluation of PV Facilities

Designing and modelling of any PV system was covered in Sect. 2. However, in reality the theoretical outcomes do not exactly match the actual outcomes, due to unforeseen conditions like (a) seasonal shading issues, (b) seasonal soiling and (c) high temperature. A few generalized case studies are described below which show the actual performance of PV systems.

Fig. 7 Change in power
output as a function of the
spectrum (AM) for PV
modules

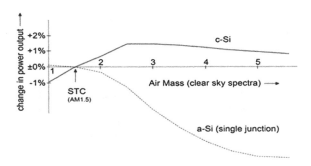

3.1 Assessment of PV Facility at Edinburgh Napier University

In April 2005, Edinburgh Napier University installed a solar PV system, costing £155,000 comprising 32 rows of BP silicon solar panels and covering a total nominal area of 160 m². Complete system was connected to the university grid. Direct current (DC) power is produced from the BP Solar high efficiency mono crystalline panels, each module produces 90 W of power at 22 V, which is then converted to a stable AC supply by four inverters. There are a total of 192 modules. The highest value of the generated peak power is 13.7 kWp (Muneer et al. 2006). Figure 8 presents a photograph of the facility under discussion.

The system contains four Fronius inverters. Two large, 6 kW (20 Amps max. current) (IG60) inverters are connected to four strings of 18 modules wired in series; each of these strings is wired in parallel with the others. Two small 2 kW (7.8 Amps max. current) (IG20) inverters are connected to two strings of 12 modules wired in the same way as the IG60 inverters. Each of the two IG60 inverters receives from the PV panels a nominal power output of 4.6 kW at 20 A. Similarly, the two small IG20 inverters receive 2 kW nominal power at 7.8 A. Each of the IG60 and IG20 inverters consumes, respectively, 12 and 7 W of electric power during their operation and 0.15 W each during the night. The AC output from the PV inverters is fed to a sub distribution board, which feeds into the building's main distribution board.

3.1.1 Measurement Facility

The PV facility is fully instrumented with both input (incident solar energy) and PV electrical energy output, recorded at a frequency of 15 min.

A. Incident solar energy

To measure the incident irradiation, two high-quality pyranometers, CM11 and CM6b, manufactured by Kipp and Zonen of The Netherlands, were used.

Fig. 8 Solar panels installed on the southeast face of the Edinburgh Napier University —Merchiston campus

B. DC and AC power output

Photovoltaic DC and AC outputs were measured via data-logging equipment installed within the inverter housing. This data-logging equipment, permits storage of the data for the whole system over long periods (up to 6 weeks at 15 min frequency) and recorded data can be viewed on a computer screen using 'Fronius-IG Access' software. It records, averaged value of AC and DC voltage, current and cumulative AC power with an energy outputs for each inverter, at 15 min intervals.

3.1.2 Local Meteorological Measurement

For the purpose of detailed analysis, it is essential to do long-term study of inclined solar irradiation. Hence, a 27-year data set of horizontal global and diffuse irradiation based on the local Meteorological Office records at Turnhouse was used to produce estimates of slope irradiation for the given PV aspect of 37° east of south and a tilt of 75° from horizontal irradiation. Figure 9 shows the long-term, hourly averaged (averaged for the whole period of 27 years) horizontal global and diffuse irradiation profile that is based on the above data set.

3.1.3 Analysis and Results

Figure 10 presents performance of the Napier PV installation on a 15 min averaged-power basis.

A. PV output calculation

Table 1 provides the calculated and measured cumulative energy quantities for the PV facility. The differences between the calculated and the measured cumulative energy are due to seasonal variations that lead to differing frequencies of cloud cover and hence the colour of the sky. An overall efficiency, $\eta = 11.5$ %, was obtained for the irradiation to AC power conversion.

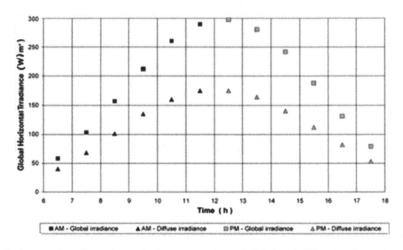

Fig. 9 Scatter plot of morning and afternoon, horizontal global and diffuse irradiance based on 27-year hourly Edinburgh irradiance data

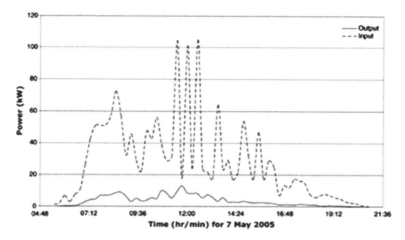

Fig. 10 PV incident power input compared with AC output for 7 May 2005

Table 1 Cumulative energy assessment for the PV installation

Month	Calculated global tilted irradiation (MWh)	Measured global tilted irradiation (MWh)	PV AC output (MWh)	PV system conversion efficiency (%)
April	10.4	Not measured	1.6	Not calculated
May	13	Not measured	1.5	Not calculated
June	12.5	10.3	1.2	11.1
July	12.6	12	1.3	11.2
August	11.6	11.2	1.3	11.9
September	8.8	9.7	1.2	12.3
October	6	5.4	0.6	10.6
Total	51.4	48.6	5.6	11.5

The AC electrical output, E was obtained from the tilted global irradiance ITLT, the useful surface area of the PV array A = 125.5 m^2 and the overall efficiency η,

$$E = I_{TLT} A \eta \qquad (2)$$

Using above equation, an analysis was undertaken to obtain the optimum PV module aspect and tilt. The chosen aspects for Edinburgh were east, southeast, south, south-west and west. For each of the aspects, the following tilts were chosen: 0° (horizontal), 10°, 20°, 30°, 35°, 40°, 45°, 50°, 60°, 75° and 90° (vertical). The result of the above analysis is presented in Fig. 11. It was found that for Edinburgh, the optimum aspect is south with the tilt range between 35° and 40°, producing the highest yield. The above yields were also compared with the Napier University installation with an aspect of 37° east of south and a tilt of 75°.

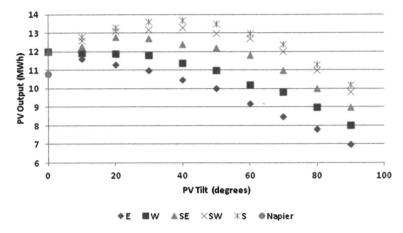

Fig. 11 Average annual PV AC output for chosen orientation and tilt

Note that, the values chosen for aspect and tilt were constrained due to architectural and planning restrictions. Hence, rather than generating 14 MWh_e AC output per annum, the PV plant is generating 11 MWh_e AC output per annum.

B. Life-cycle assessment

An LCA was performed to evaluate the embodied energy as well as the environmental impact of the installation. A list of materials that were used in constructing the given PV facility is provided in Tables 2 and 3. As a result it was

Table 2 LCA for the PV installation: embodied thermal energy

Element	Materials	Mass (kg)	Embodied energy (MWh_{th})	Co_2 released (kg)
Total spigots	Steel	200	3.3	440
Total vertical rails	Steel	2000	33	4400
Total tie brace	Steel	250	4.2	570
Total stucco rails	Steel	500	8.4	1100
Total PV module	Mixed	1500	180	3500
Cables	Copper	150	2.8	760
Total			230	11000

Table 3 LCA for the PV installation: embodied electrical energy

Element	Materials	Mass (kg)	Embodied energy (MWh_e)[a]	CO_2 released (kg)
Inverters (2 IG60)	Mixed	40	2	1700
Inverters (2 IG20)	Mixed	24	1	700
Operation			2	1100
Total			5	3500

[a]Energy imported from grid over the 30 years life of installation

Table 4 Data used for LCA calculations

Elements	Global warming potential (g CO_2/kg)	Energy coefficient (MJ/kg)
Steel galvanized from ore	2200	60
PV module silicone m-c	2400	430
Inverters (2 IG60)	42000	200
Inverters (2 IG20)	28000	140
Cables	5200	70

found that Napier University PV installation has 230 MWh$_{th}$ and 5 MWh$_e$ of embodied energy and this is equivalent to 14.5 metric tons of CO_2 emission.

Further, to calculate energy payback time (EPTB), Table 4 is used, based on data obtained from several sources. The embodied EPTB is defined as the ratio of embodied energy, converted to electrical energy and the annual electrical energy production.

$$\text{EPBT} = \frac{E_{\text{EMB}}\eta_{\text{TH}-E} + E_{\text{EMB},E}}{E_{\text{GEN},E}} \tag{3}$$

The EPTB for the Napier PV facility in the present study is estimated as ~ 8 years. This payback period is compared with other similar installations in Table 5.

In addition, Fig. 12 shows the energy payback scenarios with respect to PV facade aspect and tilt variation.

Note that if the aspect and tilt were set to optimum, then the payback times would reduce quite sharply to just over 6 years. Hence, it is always necessary to use correct aspect and tilt while designing any solar PV system.

3.2 Assessment of PV Facility in Turkey

This case study (Aldali et al. 2013) provides an experimental verification of conversion of solar irradiation from horizontal to sloped surfaces, photovoltaic cell temperature and an analysis of photovoltaic conversion efficiency. This experimental setup was installed in Southern Turkey. In addition, to measuring current and voltage of the photovoltaic module, experimental variables such as ambient temperature and solar irradiance were measured and used for validation.

Table 5 Energy payback equations

Installation	EPBT (years)	Incident solar irradiation (kWh/m^2/year)
Kannan et al. (2006)	5.9	1635
Kato et al. (1997)	15.5	1427
Present study	8	800

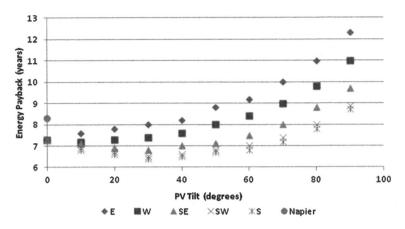

Fig. 12 Energy payback for chosen orientation and tilt

3.2.1 Photovoltaic Module and Experimental Setup

The experimental system consisted of 120 W mono crystalline photovoltaic modules situated at the top of building in Iskenderun (36°35′13″N; 36°10′24″E). The modules were tilted at an angle equal to the latitude of location, facing south. The technical characteristic of the photovoltaic modules are: short circuit current $I_{sc, ref}$ = 7.7 A, open circuit voltage at a reference condition of $V_{mp,ref}$ = 16.9 V and power at maximum power point, $P_{mp,ref}$ = 120 W. These modules are made up of 36 cells connected in series, each of 0.027 m^2 area, adding to a total of 0.974 m^2. The electron band gap, E_q = 1.124 eV. The layout of experimental system is presented in Fig. 13.

In addition, to the modules, the experimental system consists of 200 Ah/12 V lead acid battery and an inverter of 600 W/12 V and 230 V/50 Hz, with a maximum efficiency of 98 %. The load connected to the system is a number of light bulbs, whose power varies between 0 to 200 W depending on the state of charge of the

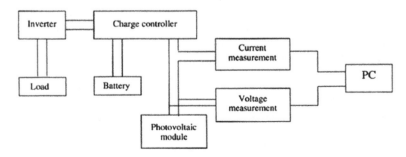

Fig. 13 Layout of the experimental setup

battery. The PV modules were connected to the battery and the load through an inverter and a charge controller.

3.2.2 Data Gathering

Four K-type thermocouples were installed and fixed on the back surface of the PV module by thermal tape to measure cell temperature and one thermocouple was installed to monitor the ambient temperature.

Long-term wind speed data was available from NASA renewable energy resource website ("surface meteorology and Solar Energy").

To measure PV performance, a HOBO silicon Pyranometer Smart Sensor with an accuracy of ±5 % was used.

3.2.3 Results and Discussion

Solar radiation data was measured with an average albedo of the ground $\rho = 0.34$ and Fig. 14 shows the comparison between actual and measured slope radiation.

It is a well known fact that, the efficiency of PV cells decreases with an increase in solar cell temperature. Cell temperature influences the I-V characteristic and therefore the electrical efficiency of the PV module. Hence, theoretical cell temperature is compared with the measured cell temperature in Fig. 15.

Finally, actual efficiency was calculated from the measured current and voltage, which is shown in terms of frequency histogram in Fig. 16. It is important to note that, efficiency values are averages of hourly data measured from June 2004 to July 2004 and it is clearly noticed that, approximately 90 % of module efficiency values are between 10 and 11 %.

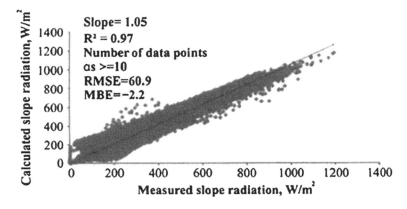

Fig. 14 Comparison between actual and measured slope radiation

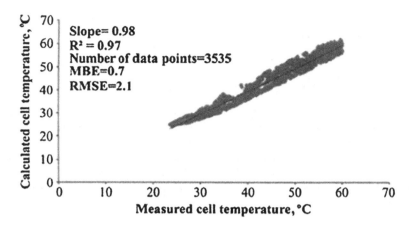

Fig. 15 Comparison between actual and measured temperature

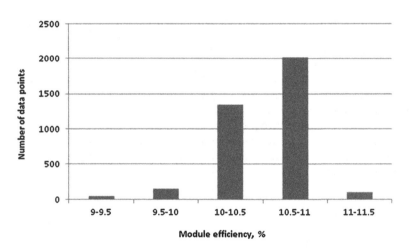

Fig. 16 Module efficiency frequency histogram

3.3 Assessment of PV Facility at Cardiff University

PV system monitoring is not universally adopted by system owners due to many reasons, including inaccessibility of the inverter display i.e. located in loft spaces or garages, cost of wireless monitoring devices and unknown weekly/monthly energy generation exception. Hence, this project was carried out at Cardiff University, Wales to see the actual performance of rooftop PV system (Sweet 2013).

3.3.1 PV System Design

By a precise architectural survey with a theodolite, it was found that roof had an aspect of 5° South of South East and the roof inclination angle was 38° from horizontal. The size of PV system fitted was 2.45 kWp. In addition, it was designed around a pre-existing evacuated tube solar thermal system in an unshaded roof area of 16.42 m².

Ten frame mounted 245 Wp mono crystalline PV modules were connected in series to one frequency transformer string inverter. Maximum Power Point Tracking (MPPT) was utilized to enable the system to react quickly to rapid changes in incident solar radiation.

3.3.2 PV System Monitoring

The inverter converts DC generated from the PV array to Alternating Current (AC); whilst simultaneously logging power data is sent in 10 min interval via Bluetooth wireless communication to a netbook runny Sunny Explorer.

The work program evaluated three software packages—PVSOL Epert v6.0, PV Watts (linked to PV outputs) and PVGIS. Results are presented below.

3.3.3 Result and Discussion

The complete monthly, annual and per cent error between actual and calculated energy (kWh) is shown in Table 6.

Table 6 Monitored and simulated energy generation (kWh)

Year 2012	System, kWh (Actual)	PVSOL, kWh (Predicted)	PV Watts, kWh (Predicted)	PVGIS, kWh (Predicted)
Jan	66.5	69.003	85	63.9
Feb	110.5	96.097	100	101
Mar	219.8	169.397	152	162
Apr	225.9	214.247	232	237
May	320.1	285.808	270	272
Jun	243.3	279.937	246	261
July	290.2	268.026	265	275
Aug	231.9	239.534	260	242
Sep	218.4	195.393	202	187
Oct	145.5	141.145	154	123
Nov	75.1	75.226	92	75.3
Dec	65	55.022	53	48.1
Total	2212.1	2088.8	2111	2047.3
Error		−5.6 %	−4.6 %	−7.4 %

Month	PVSOL	PV Watts	PVGIS
January	3	29	−3
February	−12	−10	−10
March	−22	−30	−25
April	3	3	3
May	−10	−15	−8
June	13	1	7
July	−8.5	−9	−8
August	3	11	4
September	−10	−7	−12
October	−2	8	−2
November	0	22	0
December	−15	−19	−29

Table 7 Percent error between case study and simulated monthly energy generation

The simulation software packages investigated gave annual energy generation figures which were lower than the actual generation from the system. PV Watts gave the lowest overall error of −4.6 %, PVSOL with −5.6 % and PVGIS −7.4 %. The annual yield of the case study was 903 kWh/kWp. The percentage errors between case study and simulated data, calculated on a monthly basis are shown in Table 7.

The study was analyzed based on solar resources, which is quite difficult due to variable resource in the UK (unpredicted weather patterns and high proportion of diffuse scattered sunlight). Mostly it is seen that, solar resource reaches maximum during June. However, the highest daily energy generation actually occurred in May (16.2 kWh). December generated the least amount of energy (65 kWh), predicted in all simulations (with significant errors ranging from −15 to 26 %) due to low solar resource of 6.5 h/day of sunlight.

High level of insolation in spring season, i.e. from March to May is beneficial for PV generating, as an added benefit, external temperature is relatively low. The m-Si PV modules have negative temperature coefficient (Voc is reduced to −0.33 %/°C; Isc is increased to +0.03 %/°C).

3.3.4 Assessment of PV Facility at Edinburgh College

This plant was constructed at Edinburgh College, Midlothian Campus near (Edinburgh), UK (Kelly 2013). Scottish and Southern (SSE) quoted the system cost for the entire project as £879,356, while the annual generation figure given was 568,611 kWh. A photo image, is shown in Fig. 17.

Fig. 17 Computer-generated image of completed site (SSE 2013)

3.3.5 Plant Specification

A total of 2,560 CSUN 245-60 P photovoltaic poly crystalline modules are used for the plant, which were produced by China Sunergy. Nominal power output of each module is 245 W with a cell efficiency of 16.78 %, which consist of 60 photovoltaic cells per module. Apart from PV modules, Power One Aurora Trio inverters were used with a maximum power input of 20 kW each and nominal efficiency of 98 %. Each inverter has two MPPT inputs, which ensure that the power conversion efficiency of the system is as high as possible, regardless of the amount of solar irradiation.

3.3.6 Experimental Site

The complete evaluation of performance was made with the help of different equipment like pyranometers, flux sensors and thermocouples and data loggers. The complete setup is described in this section. In addition, the circle shows the exact location of the experimental setup in Fig. 18.

A. Pyranometers

Three pyranometers were installed between two modules, in a group of modules shown as blue in Fig. 19.

B. Flux sensor and thermocouple setup

As shown in Fig. 20, four flux sensors and two thermocouples were installed on the back of the solar module.

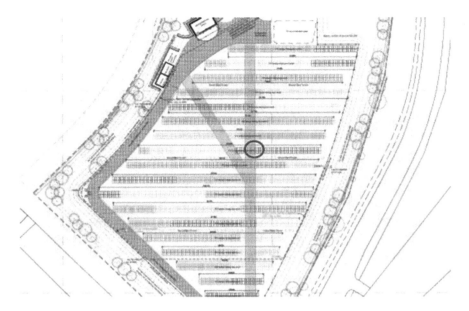

Fig. 18 Site string plan (Archial 2013)

C. Data loggers

Two, Grant 2020 series 'Squirrel' data loggers were used in this experiment as shown in Fig. 21. One data logger was used for flux sensors and the other for the thermocouples and pyranometers. They were kept in metal storage box as shown in Fig. 22, which was bound to the solar panel framing. To allow the sensors and wires to enter the box, a hole was drilled in the side of the box.

3.3.7   Results and Discussion

A. Solar irradiation data readings

Figure 23 shows horizontal (thick line) and slope (thin line) readings for a single day. A number of characteristics may be observed which correspond to expectations. With heavy cloud, horizontal and slope readings are very similar. This is due to the low levels of beam irradiation rendering the angle of incidence between the sun vector and a normal to the collector plane less important, while the horizontal pyranometer picks up slightly more diffuse irradiation, being exposed to the whole hemisphere of the sky.

During sunny periods, the slope pyranometer gives significantly higher readings for much of the day, as it is facing the sun with a lower angle of separation. This is in fact the main reason for positioning solar modules on a tilt. In the evening, this advantage decreases until the horizontal sensor is actually picking up slightly more

Fig. 19 Pyranometer
Setup. The location of
pyranometers is marked in
boxes. Note that the
pyranometer 'p1' measures
horizontal- while 'p2' and
'p3' recorded slope irradiation

beam irradiation in the hour or so before sunset (when the sun is slightly to the north).

There were no possible sources of shading on the horizontal detector and no inconsistencies observed in the recorded data. It is inferred that the horizontal irradiation readings are accurate.

B. Air and cell temperature readings

The temperature readings were recorded using thermocouple sensors denoted as 't1', 't2', 't3' and 't4'. The first two sensors measured the cell temperature from the back of panel and the remaining two were measuring air temperature. Figure 24 indicates these readings as t1 and t2 (thin-) while t3 and t4 are shown as thick lines.

C. Heat flux readings

Figure 25 displays typical characteristics to be observed across the range of recorded days. The heat flux sensors were titled 'flux1', 'flux2', 'flux3' and 'flux4'. Flux sensors 2 and 4 were found to provide erroneous signal (horizontal lines within the plot) with no change of output during the entire period of recording.

Fig. 20 Location of flux and
temperature sensors. The
boxes indicate the position of
flux and temperature sensors

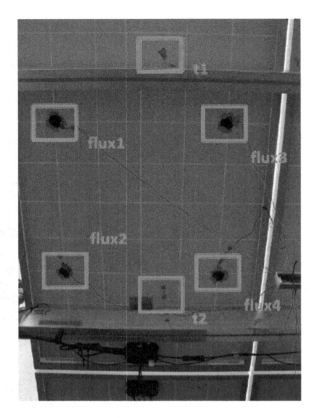

D. Cell temperature

The cell temperature was again calculated using a range of methods, the simple but widely used NOCT model, a more complex one that is used in HOMER software, and finally the full, iterative thermal model (Aldali et al. 2011). It was expected that each would give successively better result, with the thermal model being significantly more accurate than the other two due to the consideration of a greater number of factors. Figures 26, 27, 28, 29 shows the results, i.e. as 'period1' data was only for 2 days, it was neglected and only 'period 2' calculation were considered, which also gives much better results.

Finally, it was concluded that, the prediction of a solar photovoltaic module's cell temperature from environmental data such as air temperature and solar irradiation proved fairly accurate and reliable, across all three different calculation methods used. The simplest method, 'nominal operating cell temperature' of the module actually gave results almost as good as the much more complex thermal model, indicating that this is a reliable and useful method to use in future.

Fig. 21 Data loggers (Grant
Instruments 2011)

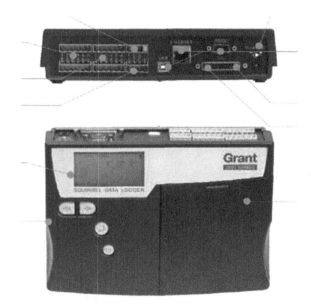

Fig. 22 Data logger housing

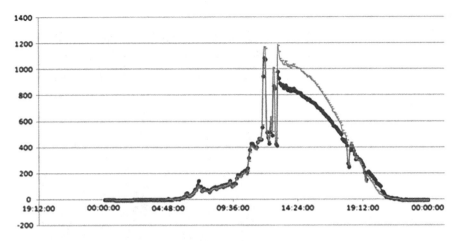

Fig. 23 Example slope irradiation measurements (W/m²)

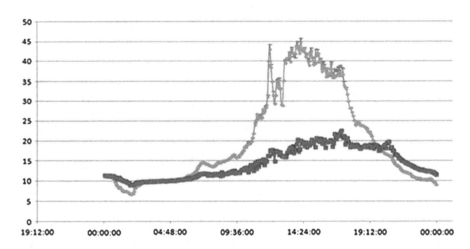

Fig. 24 Thermocouple temperature readings (°C)

4 Off-Grid Electricity System at Isle of Eigg (Scotland)

4.1 Introduction

Isle of Eigg is the second largest island of the small Isles Archipelago in the Scottish Inner Hebrides. It is located at 56.9° North latitude and 6.1° West longitude. It lies about 20 km (12 miles) off Scottish west coast, south of the Isle of Skye.

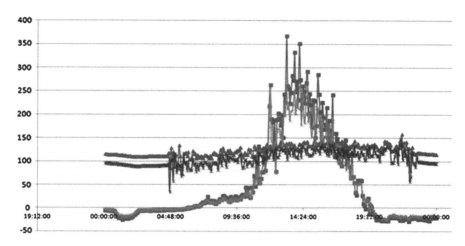

Fig. 25 Example heat flux readings (W/m²)

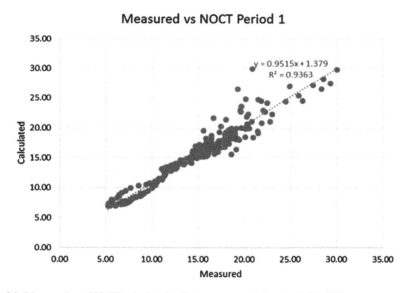

Fig. 26 Measured vs. NOCT-calculated cell temperature during Period 1 (°C)

4.2 Off-Grid Electricity System

The remoteness of Isle of Eigg from the Scottish mainland has proven to be uneconomical to connect to the national grid. Hence, in 2004 it was decided to develop a hybrid electrification system in which system load will be distributed

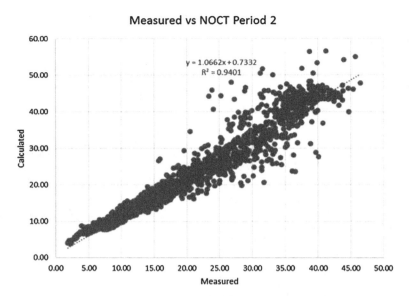

Fig. 27 Measured versus NOCT-calculated cell temperature during Period 2 (°C)

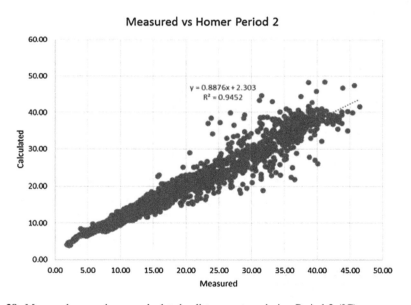

Fig. 28 Measured versus homer-calculated cell temperature during Period 2 (°C)

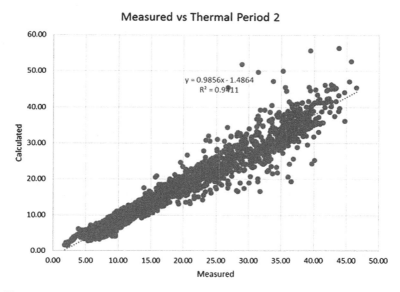

Fig. 29 Measured versus thermal model-calculated cell temperature during Period 2 (°C)

between 38 households and five commercial properties, connected through an 11 km long underground high voltage distribution system.

Chapter 1 describes the basic details of off-grid applications. However, the off-grid system of Isle of Eigg is made up of the following: 119 kW of hydro power capacity using three turbines of 100, 10 and 9 kW at three sites, 24 kW of wind power capacity (4 × 6 kW), about 54 kW of solar PV capacity and 160 kW of diesel generator capacity as back-up (2 × 80 kW).The total system installed capacity is about 357 kW. Further guidelines for selection of solar PV components can be for off-grid application can be obtained from Chap. 4.

The battery bank and the inverter system are at the heart of the system. A 48 V battery with 4400 Ah capacity [4] provides enough storage from renewable energy sources for delivery when the demand arises. The inverters control the frequency and voltage of the grid balancing the demand and supply and controlling power input and output to and from the batteries.

4.3 Investment, Timeline and Management

It took almost 4 years and the investment of £1.66 million, to put the system in place. The Eigg Electric Ltd., a subsidiary of the Isle of Eigg Heritage Trust, manages the electricity supply and distribution activity on the island. Figure 30

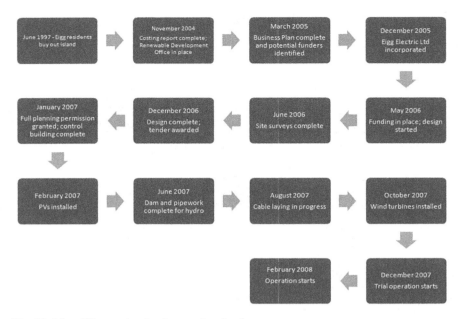

Fig. 30 Isle of Eigg project implementation timeline

details the project timeline. In routine, simple day-to-day maintenance is carried out by local residents but professional maintenance work is done by Scottish Hydro Constructing.

4.4 Simulation of Load Assessment

Load assessment simulation was carried out with the help of HOMER software. It is an optimization tool that is used to decide the system configuration for decentralized systems through simulation using different system configurations when necessary input data is supplied. It creates a list of feasible system designs and sorts the list by cost effectiveness that helps to make a decision about the optimal system configuration. In addition, it comes with a large database of technologies that facilitates the component selection process.

HOMER analysed that during winter months the average power consumption is 43 kW, while for the rest of the year, the demand is 35 kW. In addition, to prevent overloading, household electricity use was capped at 5 kW and for commercial properties at 10 kW each. It is also interesting to note that although the islanders have a capped load of 5 kW, on average they consume just 20 % of the maximum allowable total load.

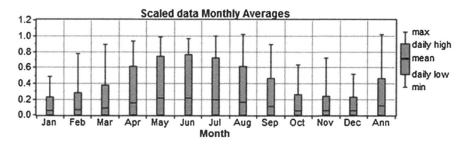

Fig. 31 Solar irradiation at Isle of Eigg

4.5 Resource Assessment

Considering the aim of this chapter, rather than assessing all components of the installed off-grid system, this section only describes the assessment of installed solar PV panels.

Figure 31 shows the HOMER sourced solar irradiation data from the NASA website. The average scaled annual solar radiation is 2.79 kWh/m²/day. HOMER also calculates clearness index which is the amount of global solar radiation on the surface of the earth divided by the extra-terrestrial radiation at the top of the atmosphere.

The total installed PV capacity at the site is 53.4 kWp, distributed in three arrays. Because HOMER does not have an option for adding panels in different time periods, it is assumed that all panels were installed at the same time and the total cost of replacement is based on current prices. According to the panel's sellers' website, the panels' life was considered as 25 years and a de-rating factor of 85 % is used for calculation. Slope of the arrays was set at the angle of site's latitude, i.e. in order to capture the most of sun energy unless the panels are equipped with tracking system and the azimuth was set as longitude of the island.

4.6 Result and Discussion

There are two load cases in this scenario, (a) for demand of 856 kWh/day and (b) for demand of 1000 kWh/day. In both the cases the energy generated by PV is 26,508 kWh/year. However, breakdown of energy production from March 2012 to March 2013 is shown in Fig. 32.

The cost of electricity (COE) is calculated to be £0.2/kWh for 856 kWh/day demand and £0.212/kWh for 1000 kWh/day demand. In addition, consumers have to pay a standing charge of £0.12 for a 5 kW load and £0.15 for a 10 kW load and initial connection charge of £500 for a 5 kW load and £1000 for a 10 kW load. Figure 33 shows the monthly production of each installed source.

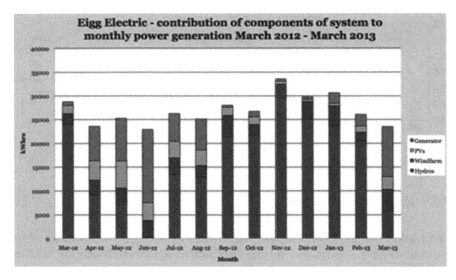

Fig. 32 Breakdown of energy production for each month

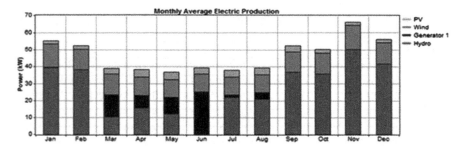

Fig. 33 Monthly average electricity production in Isle of Eigg system

Overall, the islanders receive 24 h reliable supply and have no complains about the supply. This demonstrates that, off-grid applications can be permanent solution for non-electrified areas. With 90 % of electricity generated from renewable electricity and the back-up system being used occasionally, it was noted that the CO_2 emission per household in the island is 20 % lower than the rest of the UK.

5 Conclusion

The main focus of this chapter was the modelling procedures that are currently available to determine the short- and long-term performance of PV systems. The steps that are included within the modelling of the complete systems are:

a- estimation of instantaneous or hourly horizontal beam, sky- and ground-reflected radiation once the horizontal global (total) radiation is provided, b-slope radiation, c-cell temperature (which may be obtained using simple, single-step or more sophisticated, iterative procedures that undertake a complete energy balance), d- cell and module efficiency and e-power output. It was shown that the single-step, simple as well as iterative procedures produce satisfactory results when compared to measured data.

Four case studies that covered a wide range of PV plant capacities were also provided. The range of the plant capacities varies from a few Watt to hundreds of kilowatt. Three standard, industry-based design software were inter-compared regarding performance assessment which was then compared with measured data set from a roof top installation. Once again it was concluded that the software performed with reasonable accuracy, albeit their predictions being short of actual, measured performance over a long term.

References

Aldali, Y., Henderson, D., & Muneer, T. (2011). A 50 MW very large-scale photovoltaic power plant for AL-Kufra, Libya: Energetic, economic and environmental impact analysis. *International Journal of Low-Carbon Technologies*, pp. 1–17. doi: 10.1093/ijlct/ctr015

Aldali, Y., Celik, A. N., & Muneer, T. (2013). Modelling and experimental verification of solar radiation on a sloped surface, photovoltaic cell temperature, and photovoltaic efficiency. *Journal of Energy Engineering, 139*(1), 8–11.

Archial (2013). Archial architects. http://www.archialgroup.com/.

Grant Instruments (2011). Squirrel 2020 Series. http://www.grantinstruments.com/media/2832/squirrel_2020_data_sheets_june_2011.pdf.

Kannan, R., Leong, K. C., Osman, R., Ho, H. K., & Tso, C. P. (2006) Life cycle assessment study of solar PV systems: an example of a 2.7 kWp distributed solar PV system in Singapore. Sol. Energy (in press).

Kato, K., Murato, A., & Sakuta, K. (1997). An evaluation on the life cycle of photovoltaic energy systems considering production energy of off-grade silicon. *Solar Energy Materials and Solar Cells, 47*, 95–100.

Kelly, I. (2013). Analysis of solar modelling techniques through experiment on a 620 kWp solar power plant at Dalkeith, Scotland.

Markvart, T. (2000). *Solar electricity*. Chichester: Wiley.

Messenger, R. A., & J. Ventre (2004). *Photovoltaic systems engineering*. Boca Raton: CRC Press.

NASA Surface meteorology and solar energy (2014). http://eosweb.larc.nasa.gov/cgi-bin/sse/retscreen.cgi?email=rets@nrcan.gc.ca.

Muneer, T. (2004). *Solar radiation and daylight models*. Oxford: Elsevier.

Muneer, T., et al. (2006). Life Cycle assessment of a medium-sized photovoltaic facility at a high latitude location. Part A. *Journal of Power and Energy, 220*, 517–524.

Muneer, T., Etxebarria, S., & Gago, E. (2014). Monthly averaged-hourly solar diffuse radiation models for the UK. *BSER&T, 35*, 573–584.

SSE (2013). SSE. http://www.sse.co.uk/.

Sweet, T., Sweet, C., Wu, J., Drysdale, B., & Jenkins, N. (2013). Case study vs simulation data for a grid-connected 2.45kWp PV system. Presented at the 9th Photovoltaic Science Application and Technology (PVSAT-9) Conference, Swansea, Wales, 10–12 April 2013.

TuTiempo (2014). http://www.tutiempo.net/en/.

Economics and Management of Off-Grid Solar PV System

K. Rahul Sharma, Debajit Palit and P.R. Krithika

Abstract Decentralized electricity systems, especially solar PV mini-grids and off-grid systems have the potential to significantly enhance the standard of living of communities in off-grid areas. Communities living in remote and off-grid locations are often characterized by low access to resources and infrastructure facilities in general, besides the lack of access to electricity. Therefore, it is also observed that quite often, such communities have limited paying capacities and lower energy demand at the outset compared to communities where the grid has been extended. It is therefore important to think carefully about the ways in which end-users of off-grid solar PV pay for the services they are procuring from these systems. In some cases end-users may purchase electricity in kilo-Watt-hours (kWh), and in others, services, in the form of a certain number of hours of light on a daily basis, for example. While standard practices for estimating how much a user should pay for electricity involve the calculation of a Cost of Generation (CoG) and then the tariff, a large number of off-grid projects use different metrics that are not based on the kWhs sold, but rather based on the type and duration of services being provided. Practically, such a system has advantages owing to simpler and lower cost of transactions and the fact that in many cases the transaction mimics those that rural consumers are already familiar with. This chapter aims to provide some insights into the estimation of CoG, and compares this standard methodology of fixing tariffs with an alternative service-based approach. Through the chapter readers will be introduced to basic concepts in the financial analysis of off-grid PV systems, a step-by-step procedure for estimating the CoG and finally a set of case studies to illustrate the value of service-based tariff setting.

K.R. Sharma (✉)
Centre for Policy Research, Dharma Marg, Chanakyapuri, New Delhi 110021, India
e-mail: rahulsharma188@gmail.com

D. Palit · P.R. Krithika
The Energy and Resources Institute, IHC Complex, Lodhi Road, New Delhi 110003, India
e-mail: debajitp@teri.res.in

P.R. Krithika
e-mail: krithika@teri.res.in

© Springer International Publishing Switzerland 2016
P. Mohanty et al. (eds.), *Solar Photovoltaic System Applications*,
Green Energy and Technology, DOI 10.1007/978-3-319-14663-8_6

1 Introduction

Solar photovoltaic (PV) serves as an ideal solution for off-grid power[1] owing to their modular nature. As discussed in Chap. 3, a variety of configurations, from 1 W LED solar lanterns to 10–100 W home lighting systems to kilo-Watt scale power plant and mini-grids can be designed for off-grid areas, depending on the suitability of the configuration to consumer needs. This has been one of the driving factors behind the large-scale penetration of solar-based solutions in many countries in South Asia and Africa where the need for low cost and user-friendly off-grid solutions exists (Palit 2013).

However, while internationally the costs of solar PV systems have been reducing steadily (OECD/IEA 2014), the system costs are still high for rural consumers in scenarios where there is little support from subsidies, and access to low cost loans from rural financial institutions is limited. Further, depending on the location, the business environment for off-grid systems deployment may not always be conducive (Bhattacharyya 2014); however, innovative business models have paved the way for greater consumer adoption and have also led to the entry of several private companies in the solar PV market (Krithika and Palit 2013). These innovations have been both technical (in terms of efficiency enhancement and cost reduction), and commercial. On the commercial side companies are learning to operate taking advantage of enabling polices and regulatory architecture wherever available and in case they are not available, these companies are developing innovative means to promote off-grid solar technologies using crowd financing or grant funds from different agencies.

Sufficient literature is available on the economic and financial analysis of renewable energy systems (Rabl and Fusaro 2001; Short et al. 2005). Economic analysis of such systems would include the internal rate of return estimation, impacts of subsidies and taxes, comparison with the benchmark costs of fossil fuels, and factors which aim to quantify environmental and health benefits, in order to arrive at a true cost of providing power from renewable energy sources. A financial analysis on the other hand would primarily focus on cash inflows and outflows considering the actual system costs, tariffs and the impact of government policies. This chapter aims at highlighting some of the techno-economic decisions behind successful models for off-grid solar PV and the considerations that project developers may incorporate in designing projects for markets where paying capacities may be low and consumers are dispersed and have limited energy demands. The attempt here is to create a bridge between the financial aspects of the project from the investor's point of view and the economic cost–benefits that can accrue to the users of such systems.

This chapter aims to achieve two objectives; one, to demonstrate a methodology for calculation of the cost of generation of electricity for an off-grid solar PV power

---

[1]Off-grid in this chapter encompasses individual systems as well as collective systems and mini/micro-grids working in off-grid mode.

plant (Sect. 1), and two, to showcase how business models are being developed in off-grid solar PV projects (Sects. 2 and 3). For the latter, we focus on highlighting the shift in designing revenue collection schemes from one that employs the kilo-Watt-hour (kWh) as a metric to one where service based fees are levied. While both methods are used in off-grid areas, we attempt to highlight the relative costs and benefits of using either method for designing the revenue collection mechanism and the business model as a whole. Therefore, while the first section includes a purely technical explanation of calculating the cost of generation of energy in kWh terms, the latter sections attempt to move away from the kWh metric and demonstrate the practical application of the technical tariff design process in off-grid areas.

2 Basic Concepts of Techno-Economic Analysis of Solar PV Systems

The objective of this section is to highlight the key components behind the calculation of the energy cost or unit (kWh) cost of generation from a solar power plant. The analysis has been simplified, such that project developers and practitioners may use commonly available spreadsheets (such as Excel) to compute cash inflows and outflows. While the primary objective of any financial analysis is to estimate the viability of the project within certain boundary conditions of costs and revenues, the assumptions and metrics used to define such costs may vary. That is, while kilo-Watt hours may be a useful way to understand the costs of electricity from micro-grids, other metrics such as cost per hour of illumination (Kandpal and Garg 2003) may be more applicable for solar lanterns while a monthly or weekly fixed tariff serves as a simpler tool for communication and monitoring of costs to rural consumers (especially in cases where electricity metres are not available).

The following section briefly describes some of the key components of financial analysis of solar PV systems. While a larger number of such formulae and definitions are the subject of financial analysis, this chapter focuses on a few definitions to enable simple calculations of cost of generation.

2.1 Basic Terms and Formulae in Financial Analysis

Cash Flows: Cash flows comprise of cash receipts (income/inflows/positive cash flows) and cash disbursements (expenditure/outflows/negative cash flows) over a period of time (week/month/year). Cash flows begin at time "0" and repeat at intervals that usually correspond to the interest period (in other words, if the interest is calculated annually, the cash flows will also be calculated annually). In off-grid solar PV projects, a large cash outflow corresponding to the capital expenditure of

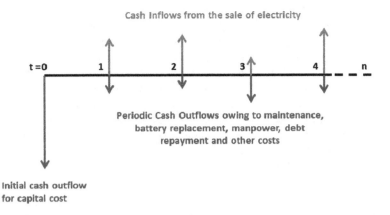

Fig. 1 Cash inflow-outflow

installing the solar power plant will generally occur at time "0", followed by annual cash inflows from sale of electricity and cash outflows representing maintenance costs, debt repayment, depreciation, periodic replacement of components such as batteries, faulty spares, human resources, etc. (Fig. 1).

Time Value of Money: The Time Value of Money is a concept which implies that one would be in a better financial situation if a certain amount of money was received today (at the earliest possible) as opposed to at a later date in the future. This is simply because money has the potential to earn interest and therefore if the money is received today, one may invest it and receive the interest amount as well in the future. For example, if in a certain scenario one can receive USD 100 today or next year, receiving it today and investing it in a savings account earning 4 % annual interest (for example) would result in USD 104 in hand next year as opposed to receiving only USD 100 next year. In addition to this, one may also factor in the impact of inflation due to which the buying power of currency changes over time and therefore the amount of goods that can be purchased decreases as the time of purchase is further in the future. In financial analysis, future cash flows have to be adjusted to account for this change in value of money. This adjustment is called 'discounting' and the factor used to make the adjustment is called the 'discount factor'.

The method by which the discount factor is calculated is a subject and is outside the scope of this chapter. Project developers may select discount factors prescribed by banks or regulatory bodies in their respective countries.[2] A popular method employed by companies for calculating the discount factor is by equating it to the

---

[2]It is important to note that as the discount factor influences cost calculations, its choice is very important. Theoretically, the discount factor should be the opportunity cost of one project relative to other potential investments.

Table 1 Use of payback period to evaluate a project

Year	Project 1		Project 2	
	Cash flows	Cumulative cash flows	Cash flows	Cumulative cash flows
0	−100	−100	-1000	−1000
1	20	−80	300	−700
2	25	−55	350	−350
3	30	−25	200	−150
4	35	10	350	200
5	40	50	400	600
Simple payback		3.71		3.43

cost of raising capital, known as the Weighted Average Cost of Capital (WACC).[3] For each year, the value of money is discounted using the following formula where n refers to the year, d to the discount factor (in percentage), Cn to the cash flow under consideration in year n and Cp to the value of Cn in present terms (i.e. year 0).

$$Cp = \frac{Cn}{(1+d)^n}$$

This procedure of including the time value of money is extremely important in solar PV projects owing to their lifetime of about 20 years, a long length of time which has large impacts on the value of cash inflows and outflows.

Simple and Discounted Payback Periods: The payback period is the amount of time in which the difference between expenditure and revenue accrued from the project (on a periodic basis) is equal to the initial investment in the project. A shorter payback period is therefore preferred as it indicates faster recovery of investment. For a simple payback period calculation, the interest or discount rate is not considered, while it is considered in the case of a discounted payback period.

A simple example in Table 1 illustrates the use of payback period to evaluate a project. Two projects with different cash flows, both positive and negative, are depicted. Taking project 1 for example, it is observed that positive cumulative cash flows are observed in year 4. Assuming that the cash flows for year 4 are uniformly distributed, we can calculate that the payback for such a project would be 3 + (25/35) = 3.71 years.

Similarly, Table 2 illustrates a calculation of discounted payback period for the same projects, in which the cash flow each year is multiplied by the corresponding discount factor. It is important to note the impact that discounting has on the payback period, which is now longer for the same projects as a result of the

---

[3]WACC = (Amount of Equity*Cost of Equity + Amount of Debt*Cost of Debt)/ (Amount of Equity + Amount of Debt). The Cost of Equity is typically equivalent to the Return on Equity expected by the investor and the Cost of Debt is equivalent to the interest paid on debt from banks or financial institutions.

Table 2 Calculation of discounted payback period for the projects

Year	Discount factors (dn)	Project 1			Project 2		
		cash flows (Cn)	Discounted cash flows (Cn*dn)	Cumulative discounted cash flows	Cash flows (Cn)	Discounted cash flows (Cn*dn)	Cumulative discounted cash flows
0	1	−100	−100	−100	−1000	−1000	−1000
1	0.91	20	18.2	−81.8	300	272.73	−727.3
2	0.83	25	20.7	−61.2	350	289.26	−438.0
3	0.75	30	22.5	−38.6	200	150.26	−287.8
4	0.68	35	23.9	−14.7	350	239.05	−48.7
5	0.62	40	24.8	10.1	400	248.37	199.7
Discounted payback at 10 % discount rate				4.59			4.20

application of the discount factor, which reduces the value of future payments (re-emphasizing the impact that discount rate selection has on the final result). Formulae for calculating discounted payback periods in a single step are available in standard books on finance, for which readers may refer to (Crundwell 2008).

Net Present Value: The net present value (NPV) is the difference between the present value of inflows and outflows through the life of the project. A positive NPV is an indication that a project may be considered viable for implementation while a negative NPV indicates that it will be a loss-making venture. The two projects considered in the examples above are used in Table 3 to illustrate NPV

Table 3 NPV calculation

	Year	0	1	2	3	4	5
	Discount factors (dn)	1	0.91	0.83	0.75	0.68	0.62
Project 1	Cash outflows	100					
	Cash inflows		20	25	30	35	40
	Discounted cash outflows	100	0	0	0	0	0
	Discounted cash inflows	0	18.2	20.7	22.5	23.9	24.8
	Sum of discounted cash outflows (A)	100	NPV (B-A) 10.12				
	Sum of discounted cash inflows (B)	110.12					
Project 2	Cash outflows	1000					
	Cash inflows		300	350	200	200	200
	Discounted cash outflows	1000	0	0	0	0	0
	Discounted cash inflows	0	272.7	289.3	150.3	136.6	124.2
	Sum of discounted Cash Outflows (A)	1000	NPV (B-A) −26.97				
	Sum of discounted Cash inflows (B)	973.03					

calculation. However, this time for Project 2, the final two cash inflows have been reduced to INR 200 per year for the purpose of illustration.

We see here that while Project 1 has a positive NPV and thus is a viable project, Project 2 has a negative NPV and will be considered as a loss-making project. Here also, standard formulae to facilitate a single step calculation are available (Crundwell 2008).

It should be noted here that keeping all other factors constant, the NPV is sensitive to selection of the discount rate. In the example above, if the discount rate was 8 percent instead of 10 percent, the NPV of Project 2 would also be positive and we would select the project with a higher positive NPV among the two. In practical terms, when a higher discount factor is selected, it implies that the current investment in the project is lower and vice versa for a lower discount factor. Hence, a careful selection of the discount rate is important.

Internal Rate of Return (IRR)[4]: The IRR is defined as the value of interest/discount rate at which the NPV of a project just equals zero. Therefore, instead of using an absolute value of money (as with the NPV), a percentage value (the rate of return) is used to determine the viability of a project. Project developers would have a generally acceptable IRR in mind and would base their decisions on going ahead with a project based on whether the project's IRR is more than or at least equal to their desired IRR. The immediate advantage of the IRR method of project evaluation over the NPV method is that the selection of a right discount rate has no bearing on the result. Rather, the purpose of IRR calculation is precisely to calculate this rate.

Using Project 1 above as an example, we see that the NPV will be equal to zero (by trial and error[5] and inserting different values of interest rate) at a rate of 13.45 %. This then is the IRR of the project. If the developer is Project 1 has a desired IRR of 16 % (benchmarks for IRR may be different in different countries, usually return on equity varies between 14 % and above), then he/she might choose not to invest in this project. Further, equity IRR (or return on equity) is different from project IRR. Project IRR is a measure of the economic viability of the project. It considers the project costs and project revenues without considering where the money has come from (debt, equity, grant, etc.). On the other hand, the equity IRR is the measure of the rate of return of an investment to the equity investors taking into consideration the equity investment (generally at the start of the project) and the returns to the shareholders over the lifetime of the project.

Again, standard formulae are available for single-step calculation of the IRR and may be found in books on financial evaluation of projects referred to earlier. It is also important to note here that for more complex cash flows, multiple IRR values

---

[4]While the IRR will not be used in subsequent calculations of Cost of Generation, the concept has been included in order to compare it to the payback period and NPV, as all three options are commonly used in financial analysis and have their relative advantages and disadvantages.

[5]Spreadsheet users may utilize the 'Goal Seek' function wherein the value of NPV is set to zero by varying the value of interest rate.

may result. However, with off-grid solar PV projects, the cash flows are generally predictable and linear, leading to a single IRR value.

Payback period, NPV or IRR are all methods used to evaluate the viability of a project and it is up to the developer to select the method most suitable to his needs. Each method has its own pros and cons and, for example, while payback period calculations are simple, they provide information only up to the payback date and do not give the investor any information about how viable the project is based on the entire lifetime of cash-flows, which is especially important in the case of solar PV projects which have a high initial investment but benefit from zero fuel costs and low operational costs over their lifetime compared to conventional power technologies.

Similarly, while the NPV does consider lifetime costs and provides the analyst a single number, thus simplifying the process, it is heavily dependent on the selection of the discount factor and does not distinguish between a large and small investment. For example, while the NPV of two different projects may be similar, one project could begin with an investment of USD 100 and another with an investment of USD 1,000,000 and therefore the NPV does not provide us with a sense of the scale and the effort required to do the project. The IRR method avoids the selection of a discount factor but in some investment decisions (in which investment to the project are made a number of time during the project life compared to a single investment at project inception) it can lead to multiple solutions for NPV = 0, thereby creating some ambiguity in interpreting results. Readers can find further inputs on the scenarios under which either method is suitable in the references provided earlier in this chapter.

In practice, a project developer may use both NPV and IRR in analysing potential investments. While one gives you a feel for relative size the other indicates absolute size. Both NPV and IRR provide valuable information in the decision making process and in most cases, they will give the same go or no-go decision.

2.2 Computing the Cost of Generation of Electricity from an Off-Grid Solar Photovoltaic Power Plant

For an off-grid solar PV power plant to be viable, the developer must recover his investments through payments for use of electricity by consumers. These payments will depend on what the cost of generating electricity from the power plant is, in other words, the cost per kilo-Watt hour (USD/kWh).

The cost estimation is entirely based on the cash inflows and outflows explained in the previous sections. In its simplest form, the cost of generation is equal to the NPV of expenses or outflows over the life of the power plant, divided by the total number of kWh of electricity sold during the life of the power plant. Developers may also be interested in knowing the annual cost of generation from their power plant, in which case this cost is estimated as the total cost of generation in any one year divided by the electricity generated in that year. The annual cost of generation will vary from year to year owing to difference in expenses year to year, especially in years where there are larger expenses owing to battery replacements. While the annual cost of generation will provide the developer with a more accurate yearly cost of generation, the levelized cost of generation provides us with a method of evaluating the cost of generation from two different projects over their lifetime.

In this section, we will first list the various components that contribute to cash outflows and, then through a simple example, demonstrate the calculation of cost of generation of electricity. It is important to note here that while cost of generation is the minimum amount of inflow per kWh required in order to meet all costs and incorporate reasonable profits,[6] the actual tariff or electricity price charged to the consumers may be higher in order to account for a number of other factors such as unexpected expenses, collection inefficiency, if any, maintenance of healthy financial ratios every year and so on.

2.3 Components of Cash Flows for Off-Grid Solar PV Power Plants

Capital cost: The capital cost of a solar PV power plant will include cost of solar modules, mounting structure, civil construction for mounting of panels, battery bank, inverters (in case of AC solar power plants), power plant room (battery bank cum inverter room), wiring, transport, and human resources required for installation and commissioning. In addition, the capital cost will also include the power distribution network and household service connections for a solar mini-grid project. In some cases, the power distribution network may already be present or subsidized by the government and the household connection cost may be borne by the consumers. Also, subsidies and tax exemptions may be available on various power plant components. Developers may subtract such costs from the total cost to arrive at the capital cost of the project[7] which should only comprise of those costs which are directly borne by the developer.

Equity and Debt: The capital cost of the power plant (after factoring in subsidies and grants) has to be raised through money available with the developer (equity) or

---

[6]In our example here, the profits and tax considerations have been excluded and may be added into the appropriate cash outflow rows of the analysis depending on developer and country specific conditions.

[7]A differentiation has been made in this chapter between Power Plant Cost and Project Cost. The Power Plant Cost includes the cost of components of the Solar PV Power Plant including solar modules, inverters, batteries, charge controllers, power plant metering and cabling, installation and commission (including the support structure for solar modules). The project cost on the other hand includes the larger set of costs inclusive of the power plant, distribution, household wiring, civil construction required for housing the batteries and inverters, administrative costs, capacity building costs, etc.

money borrowed from financial institutions (debt). Generally, the ratio of debt to equity is 70:30 but may vary depending on the particular situation and country of implementation.

While the developer will earn a return on his equity investment (and incorporate the return on equity in the cash flows), he/she will also be paying the money borrowed back to the bank on a periodic basis, along with the interest. The rate of interest on debt would be set for a project based on prevailing interest rates in the country or the negotiated interest rate with the agency providing the debt and the element of risk in the project. It has generally been observed in developing countries that banks tend to charge higher interest rates on rural electrification project owing to the perception of a higher risk of recovery and the loan term can be quite short, in the range of 5–7 years (ESMAP 2001; IED 2013).

It is important to point out here that while there is an ongoing debate on the subject, grants in general, should not be treated as equity investments in a project. Grants are non-returnable capital, while equity is returnable capital where equity investors expect a reasonable return commensurate with their risk. However, we also find cases where grants have been treated as equity (under special conditions). For example, in Tanzania, commercial banks require the small power producers to have 30–40 % of their own equity to finance capital costs of the mini grids; however, there are very few private producers that can provide such investment from their own resources. To help close this equity gap, the banks allow the treatment of "connection grants" provided to the power producers by the Rural Electrification Agency (REA), which are to the tune of $500 per connection, to be treated as the own equity of the producer. This is mainly for the purpose of meeting the bank's minimum equity requirements. However, this is 'neutered equity' as the power producer cannot earn an equity return on this gift of equity (Simpa Networks 2014). This is treated differently from normal equity that is supplied by the power producer or an outside investor. In case of REA grant, it is provided as a gift where the REA does not expect to earn any return but expects the power producer to connect a specific number of households, whereas in the latter case, both expect to earn a return on it. Therefore, the regulatory treatment also differs in both cases. For the REA grant, the power producer is allowed to take depreciation on the capital financed by the grant, but not allowed to earn an equity return on the grant while for the normal equity both a return on the equity supplied and depreciation on any capital equipment financed by the equity is allowed.

Operation and Maintenance Cost: The operation and maintenance costs for off-grid solar PV projects would comprise of human resources costs (say, a local operator for everyday operation, maintenance, weekly cleaning of solar modules, weekly/monthly collection of revenue, etc.), spare parts and replacement of components such as batteries and inverters, which have a shorter life than the 20-year life of the solar panels. In case of lead-acid batteries for example, field-experiences show that the life of a well-maintained battery is in the range of 5–7 years. While batteries with longer lives such as lithium ion batteries are available, their cost will be higher.

While these maintenance costs may be calculated by summing up all the above mentioned costs, as a thumb rule, the maintenance cost for off-grid solar PV power plants may be assumed to be about 3–5 % of the capital cost of the power plant on an annual basis.

Depreciation: Depreciation is the decrease in the value of an asset (such as solar modules of a solar power plant) over its useful life. Depreciation is a characteristic of all physical assets and is caused by a combination of factors such as normal wear and tear, deterioration and obsolescence. The different methods of depreciation (two common methods are Straight Line Method and Written Down Value Method) charging or terminologies such as book value, salvage value, etc., are not covered in this section and readers will find such information in the references mentioned earlier in this chapter (Crundwell 2008). However, for the purpose of this chapter, the straight line depreciation method is used, wherein the value of an asset decreases linearly with respect to time. The depreciation rates for different types of equipment are also generally specified by regulatory bodies. For example, with respect to India, the depreciation rate specified by the Central Electricity Regulatory Commission (Central Electricity Regulatory Commission 2014) (CERC) is 5.83 % for the first 12 years of the plant life and 1.54 % from the 13th year for a solar power plant which is used here for calculation.

Insurance, Taxes and business development expenses: These costs will be specific to the country where the project is being planned and need to be factored in along with other cash outflows.

Revenue: The revenue for off-grid power plants will be from the sale of electricity to different consumers. A few points to be kept in mind while calculating the revenue are as follows:

- *What will be the annual production of electricity?* For example, in many cases in India, it is assumed that an off-grid solar PV power plant of 1 kWp capacity will generate 4 kWh of electricity per day and will be in operation for 300 days of the year. Based on this, one may assume that a 1 kWp power plant will produce 1200 kWh of electricity per year. This average value is considered for our example here, and is derived from the interactions with a large number of developers of small-scale solar PV power plants in India
- *How much electricity will actually be consumed?* It is highly unlikely that all the power that can be produced in a solar power plant will be consumed at the beginning of a project. Some time will elapse before consumers get new connections or loads build up and since the provision of feeding excess power to the grid does not exist in off-grid power plants, not all the electricity that can be generated will be used. Hence, realistic projections of load growth should be built in to avoid an overestimation of revenues and for optimal generation of electricity from a power plant. In a second scenario which has been observed frequently in the field is that the solar PV power plant may be overdesigned, thereby increasing the capital cost and therefore the unit cost of generation. Hence, it is important to design the system optimally and match load with the generation.
- *Will the power plant produce the same amount of power each year?* For any system, some amount of degradation in output should be assumed. For solar PV

power plants, module manufacturers specify this annual degradation rate (generally, 10 % over the first 10 years of operation and 20 % over the remaining 10 years of operation). Even within one year, the power generation will vary from season to season. In the case of India for example, the power generation drops significantly during the monsoon season or during winter foggy situation. Hence, it is also important to note and design the power plant as per the geography and climatic conditions in which it is operating.

- *Will all consumers purchase power at the same tariff?* It is likely that the power plant will have domestic, agricultural and commercial consumers who can pay different rates of tariff. This should be factored in assuming a certain fraction of each kind of consumer. Such a differentiation in consumers provides the developer the opportunity to connect larger and more predictable loads from agricultural pumping and small and medium-scale enterprises. However, in the case of off-grid sites, it is also likely that all the consumers will be households or the project developer may make an informed decision to only supply power to a certain consumer segment in which differentiation in tariff will not be required.

- *Is the tariff set on a kWh basis or some other metric?* If a service-based tariff has been employed, then rather calculating the cash inflow as a product of electricity sold in kWh and the tariff per kWh, the inflow would be calculated as the product of the number of connections and the charge per connection over a period (day, week, month or year). This type of tariff setting will be discussed in detail in Sect. 2.

2.4 A Simple Illustrative Example

Using the concepts discussed above, an example of how the tariff (per kWh) is calculated is shown below. The case study is of a 10 kWp solar PV power plant and some of the key parameters are depicted in Table 4 below:

Using these parameters as inputs, the cost of generation in Rs/kWh or USD/kWh is calculated. This single figure of 13.8 Rs/kWh (0.23 USD/kWh[8]) is the fixed cost of providing electricity from this power plant for a period of 20 years. In other words, it is calculated as the Net Present Value of total expenses over the lifetime of the project, divided by the total electricity generated over the lifetime of the project. In order to arrive at a cost of supply, developers would need to increase this value by factoring in losses, insurance, profits, taxes, etc., to the expense head.

The calculations in Table 5 have been shown till year 6 only, in order to fit the example into the page. In the third column in Table 5, wherever Roman numerals have been used, the reference is to the inputs parameters in Table 4 and wherever capital letters are used, the reference is to rows within Table 5 itself.

---

[8]An exchange rate of 1 USD = 60 INR has been used throughout the chapter.

Table 4 Key input parameters of the solar PV off-grid power plant

S. No.	Input parameters	Unit	Remarks
(i)	Plant capacity (kWp)	10	(or 10,000 Wp)
(ii)	Benchmark capital cost—capex (Rs/kWp)	170,000	The benchmark capital cost has been derived from the benchmark costs issued (Ministry of New & Renewable Energy 2013) by the Ministry of New and Renewable Energy (MNRE), Government of India.
(iii)	O&M (% of capex) excluding battery change	3 %	Standard industry practice
(iv)	Annual escalation	7 %	Proxy for the rate of inflation, computed using the average of the Consumer Price Index for the past 5 years in India
(v)	Battery cost (% of capex)	25 %	Average of costs specified by local installers of PV systems
(vi)	Life of PV system	20	Standard industry practice
(vii)	No of days of Operation per year [8]	300	Standard industry practice
(viii)	Equivalent sun hours per day	4.0	Average value obtained from stakeholder consultation with off-grid Solar PV industry representatives and will vary depending on location
(ix)	Annual derate (%) in PV module efficiency	1 %	Standard industry practice
(x)	Grant/capital subsidy (% capex)	Variable %	Depending on the subsidies available locally. For our case, no subsidy or 0 % is assumed
(xi)	Equity (% of capex)	30 %	Standard industry practice
(xii)	Return on equity (%)	20 %	Standard industry practice
(xiii)	Interest rate (%)	12 %	Standard industry practice
(xiv)	Discount rate (%)	15.0 %	Standard industry practice: the weighted average cost of capital
(xv)	Loan repayment period (years)	10	May vary depending on project risk and standard banking practices
(xvi)	% Depreciable capex	90 %	Derived from the CERC guidelines
(xvii)	Annual depreciation rate—first 12 years	5.83 %	Derived from the CERC guidelines for India
(xviii)	Annual depreciation rate—from the year 13	1.54 %	Derived from the CERC guidelines for India

Users of this method will notice that while the annual cost of generation in Rs/kWh or USD/kWh is high, the discounted cost is much lower and this is a result of the discounting of costs up to the period under consideration. The graph in (Fig. 2) depicts the variation of the annual average cost and the discounted cost of

Table 5 Calculation of cost of generation

Parameter	Unit	Formula	Values	2	3	4	5	6
(A) Capital cost post subsidy	Rs	(i)*(ii)*(x)	1,700,000					
(B) Debt	Rs	(B)*(xi)	1,190,000					
(C) Equity	Rs	(B)*(1–xi)	510,000					
(D) Battery cost	Rs	(B)*(v)	425,000					
(E) Years			1	2	3	4	5	6
(F) Electricity generated	kWh	(i)*(vii)*(viii)*(ix)	12,000	11,880	11,761	11,644	11527	11,412
(G) Discount factors	No.	(1–(xiv)^(E))	1.00	0.87	0.76	0.66	0.57	0.50
(H) Interest factors	No.	(1 + (iv)^(E))	1.00	1.07	1.14	1.23	1.31	1.40
(I) Return on equity amount	Rs	(C)*(xii)	102,000	102,000	102,000	102,000	102,000	102000
(J) Interest amount	Rs	(using an interest calculator)	135,660	121,380	107,100	92,820	78,540	64,260
(K) Depreciation amount	Rs	(A)*(xvi)*(xvii)	89,199	89,199	89,199	89,199	89,199	89,199
(L) O&M cost amount	Rs	(A)*(iii)	51,000	54,570	58,390	62,477	66,851	71,530
(M) Battery replacement cost	Rs	(D)*(E)	0	0	0	0	0	557,088
(N) Total expenses	Rs	Sum (I to M)	377,859	367,149	356,689	346,496	336,590	884,077
(O) Discounted expenses	Rs	(N)*(G)	377,859	319,260	269,708	227,827	192,446	439,543
(P) Annual cost of generation	Rs/kWh	(N)/(F)	31.5	30.9	30.3	29.8	29.2	77.5
(Q) Discounted cost of generation	Rs/kWh	(O)/(F)	31.5	26.9	22.9	19.6	16.7	38.5
(R) Expenses NPV	Rs	Sum of all values in row O	3,004,663					
(S) Total electricity generated	kWh	Sum of all values in row F	218,512					
(T) Levelized cost of generation	Rs/kWh	(R)/(S)	13.8					

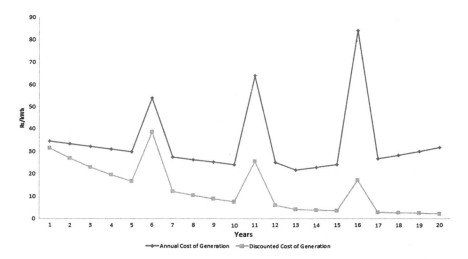

Fig. 2 Variation of the annual average cost and the discounted cost of generation over time

generation over time. The peaks observed in the graph correspond to the years when battery replacement takes place, leading to a much higher cost of generation for that year.

3 Site-Specific Considerations for Selecting System Configurations and Designing Business Models

In addition to the techno-economic analysis detailed out in the previous sections, there are a number of factors which ultimately determine the long-run viability of an off-grid electrification project, which are context specific. Such an approach primarily underscores the need to conduct a holistic assessment starting with economic and financial feasibility of projects, to consider cultural, environmental and social norms during project implementation.

Therefore, going beyond the analysis shown in the previous section, this chapter also aims to discuss some conditions that impact business models for off-grid solar PV power plants, beyond what a simple Excel sheet might be able to demonstrate.

The advantage of working with Solar PV is its modularity and that it can be designed easily for both AC and DC applications. These features offer the developer of such systems, a larger number of possibilities in designing technical and business models, depending on the conditions of the site of proposed operation. Some of these factors are discussed in this section.

3.1   Characteristics of Off-Grid Sites and Important Considerations for Selecting System Configurations and Business Models

Off-grid communities in South Asia and Africa are typically communities that do not have access or have poor access to other essential infrastructure such as good quality roads, job opportunities, communication facilities, health and education among others. This essentially translates to these communities having low paying capacities, low load demands and growth and the need to migrate. While setting up a single solar PV power plant and extending distribution lines to each household and other community buildings may be a simple option which replicates the existing grid, practice and research over the past decade has found that there are much more feasible options outside the legacy system. Before we showcase the features of such options, a few of the characteristics of such communities are discussed in this section.

Small cluster of households: In India, as per government figures (Central Electricity Authority 2011) the grid has reached 95.7 % of the population. If we accept the definition of 'connectivity' as one where electricity has reached at least 10 % of the households in a village and public places (Ministry of Power 2014), this leaves out a small number of villages and hamlets that are completely off-grid. Typically, such villages and hamlets have around a maximum of 50 households with none or few existing loads which makes the economics of setting up a solar PV power plant weak. However, more sustainable business models have been developed for solar charging stations and solar DC microgrids for the provision of lighting and mobile charging facilities until higher demand growth is observed. Similarly, even in many grid-connected areas where electricity supply is erratic (less than 6 h a day), such models have been successfully implemented.

Scattered households: In remote locations in many African countries, it has been observed that the distance between household and villages is often quite large (Sharma and Palit 2014). Covering such distances with distribution lines would lead not only to high distribution losses but also increase the cost of the overall system significantly. Again, more modular models for energy service delivery such as stand-alone systems, charging stations or micro-grids are more appropriate in such a context.

Ease of management, repair and maintenance: In many off-grid communities, the cost of involving experts to perform roles of management, repair and maintenance is often expensive owing to the remoteness of the locations. It has therefore been suggested that local skills must be enhanced in these domains (Chaurey et al. 2012). However, involving the community in technical maintenance may not be feasible beyond a point owing to the limitations of capacity building measures and the existing knowledge base within the community. In other cases, the community may also desire that a trained expert is in charge of these functions. One option to resolve this, besides long-term capacity building is in the reduction of the complexity of the system installed. For example, a solar DC microgrid, which does not involve an inverter, is technically a simpler system to maintain locally, in cases where the community's energy needs are limited. Generally, simple electrical equipment such as cables, bulbs, fuse, battery charge controllers and so on are available in nearby towns, as opposed to solar power plant inverters which are more expensive and complex. Developers may therefore also consider this reduction in system complexity while designing systems for remote locations.

Ease of transactions and extent of awareness within the community: The concept of units or kWh of electricity is complex and at times not well understood even by many city dwellers who have been using electricity for a long period of time. However, this use over time does ensure that they are well accustomed to the costs of power and payment mechanisms, regardless of their familiarity with electricity itself. In off-grid communities that have never used electricity or are familiar only with options such as fee-based battery or mobile phone charging, it is at times useful to communicate the charges of using electricity in service based rather than kWh-based terms.

This is especially relevant in cases where only simple services such as lighting or mobile phone charging are being provided. It is also easier for communities to compare costs, say on a monthly flat-fee basis with their existing costs of purchase of kerosene. For example, a user who is currently spending INR 150 (USD 2.5) per month on kerosene will be immediately able to see the benefit of spending the same or lower amount of money on one light point of a higher quality.

For the project developer, such a service-based approach also has the additional benefit of excluding the costs of metering and metre maintenance from project cash flows. Instances of metre tampering and disconnection are also avoided although other forms of misuse such as overloading of the connection may need to be resolved using other technical measures such as load limiters.

Opportunities for income generation and supporting other local institutions: While rural energy systems are generally designed for households, developers may also find opportunities in supporting income and livelihood generating activities or institutions such as hospitals, schools and community buildings. These options can become a significant input to designing systems of higher capacity and operating three phase loads.

It is well understood today that the introduction of productive loads can enhance the business model by not just ensuring a constant base load for the power plant from large loads such as grinders, mills, water pumps and mobile phone towers, but also the commercial rate of tariff that may be charged to such customers (Palit et al. 2011). There are of course challenges in countries where subsidies for agricultural pumping exist, however, in most other cases, such loads can become anchor loads for the project and enhance revenue streams. This improves the possibility of designing a larger power plant providing AC electricity to all connected consumers including households.

For solar power plants, however, the inclusion of productive loads can also significantly increase the capacity and therefore installation cost of the power plant. In case of the biomass or micro-hydro power plants, the increase in installation cost of the power plant is not as high relatively owing to the economies of scale achieved with larger capacity biomass or micro-hydro power plants. On the other hand, solar PV is modular in nature and especially at the smaller scale, the cost of increase in capacity of the power plant (say by 25 %) to accommodate productive loads, for instance, will not lead to a decrease in overall cost on a unit basis as a result of scale. Therefore, while productive loads are an important means of increasing the CUF and revenues, solar PV power plants are possibly better suited to serve a larger number of smaller loads at current costs.

Tariffs—redefining 'grid parity': Beyond the ease of communicating and recording non-kWh-based tariffs, service-based models have other advantages. While a kWh-based metric is well accepted, it is also important to consider more bottom-up approaches to tariff design and find a convergence point of 'grid-parity' between grid based and off-grid tariffs. It is clear that if the tariff from an off-grid solar power plant is communicated purely in kWh terms, it may appear high and there will be few takers.

However, while comparing the tariff from a solar PV-based mini-grid to the tariff of the conventional grid, it is vital to also consider the nature of each source of electricity. The central grid integrates a large customer base, benefits from cross-subsidies and other forms of financial support which are generally not available to mini-grid operators (Bhattacharyya 2014).

On the other hand, it is also important to evaluate the coping cost of the end-users, that is, the amount of money spent on a monthly basis to cope with or find alternatives to the absence or poor functionality of the grid. This, for example, could be in the form of expenditure on kerosene or diesel or battery charging from nearby grid connected villages. Moreover, these costs are usually incurred on a weekly basis, a pattern of payment that becomes familiar to low income consumers. Therefore, good practice would also attempt to replicate this payment pattern, rather than burdening the end-user with a large monthly or bi-monthly bill (as in the case of grid power) which the end-user may not be able to pay in a single installment.

But even if we set aside the coping cost and are only concerned with the cost of electricity provision, the kWh basis of comparison may not always gives us a correct figure of what a household spends per month on an electricity connection. Electric utilities also include a component of Fixed Cost in their billing, which is independent of the number of kWh consumed per month. This fixed cost, for example, amounts to INR 180 (USD 3) per connection per month with no energy (kWh) charge for unmetered connections and 50 INR/kW/month (0.83 USD/kW/month) with 2.20 INR/kWh (0.035 USD/kWh) as energy charge for metered connections in the state of Uttar Pradesh in India, for consumers getting supply as per the 'Rural Schedule' of the electricity distribution company in the state (Uttar Pradesh Electricity Regulatory Commission 2013). When this total cost is compared to the cost of monthly solar lantern rental or DC microgrid tariff, consumers are in a position to make a more informed decision regarding the cost and quality of service rather than a comparison only involving energy or kWh costs. For example, in the case study on Mera Gao Power (MGP) covered in Sect. 3, the tariff is INR 25 (USD 0.42) per week or INR 100 (USD 1.67) per month. If unmetered consumers who receive only 6–7 h of intermittent supply from the grid for two light points compare their current grid electricity tariff to the tariff charged by MGP, it is likely that they will find MGP's tariff more suitable to their requirements.

3.2 Beyond Kilo-Watt Hours: Options for Off-Grid Customers

While kilo Watt hour provides us with a simple metric to measure the electricity services being provided to rural households (say, 1 kWh per month as the lifeline energy requirement), with the emergence of non-grid connected options, governments, donors and private players have expanded to other options such as lanterns, task lights, home lighting systems, micro-grids, etc. Such systems enhance the flexibility of providing service and lower the costs associated with extending distribution lines to areas where such extension can be expensive. From a broader policy perspective, they encourage decentralization and local management of electricity resources and reduce the burden on the legacy system to provide power to remote and underserved populations. The aim of this section is to briefly describe these options and, show through examples, the ways in which business models are being developed around these systems.

3.2.1 Solar Lanterns, Home Lighting Systems and Other Stand-alone Systems

Solar lighting systems have emerged as a viable alternative to kerosene or diesel based lighting solutions in off-grid areas when viewed from the holistic perspective of costs, health and light output. The transition from CFL-based systems to LED based systems in recent years has also aided in significantly reducing the capacity and therefore cost of the system, making these readily accessible to low income populations (Thakur et al. 2011).

Solar lighting systems may be classified into solar lanterns, solar task lights, solar home lights, solar street lights and solar torches. The price range and applications vary and while a solar lantern may provide 360° of ambient light for household activities such as cooking or eating, a task light provides more focused light for an activity such as reading. Such devices have been promoted widely through programmes such as the Lighting Global programme of the International Finance Corporation, the IDCOL Solar Home Systems programme in Bangladesh, the Lighting a Billion Lives (LaBL) programme by TERI among others.

3.2.2 Solar Charging Stations

One alternative to sale or dissemination of stand-alone solar lighting systems has been the Solar Charging Station. The Charging Station concept was widely demonstrated by TERI since the year 2008 through its Lighting a Billion Lives Programme (Palit and Singh 2011) and has then been also used in other initiatives such as the Solar Transitions Project in Kenya (Ulsrud et al. 2011). A Charging Station is designed on the premise that a single entrepreneur in a village would be responsible for maintenance of the solar lanterns and would rent them out to customers. Lanterns are connected in groups of ten or more to a solar module through a charge controller and junction box and their batteries are charged during the day. By evening, once the lanterns are fully charged, users come to the charging station, rent the lanterns and pay a small fee for the service. The lanterns are then returned the following morning. The advantage of such a model is that the repair and maintenance of the lanterns and the system as a whole remains centralized and with one entrepreneur—making the task of procuring spares and maintaining a clear schedule for maintenance simple. The drawback is that in some cases, users prefer to own their system rather than rent it on a daily basis. The lack of ownership also sometimes translates to misuse and damage to the lanterns. However, owing to the small fee of about INR 3 (USD 0.05) per day in India, such a model has found a large number of takers in areas where paying capacities are low and opportunities, such as monthly installments or assistance from financial institutions do not exist for the purchase of a complete stand-alone lantern.

In the Solar Transitions Project, implemented by a research consortium led by University of Oslo, the Charging Station not only supports lantern charging and rental but also other activities such as typing, photocopying, printing, television viewing, hair trimming, etc. These activities require small amounts of power only and the entire charging station has been designed for 2.16 kWp. Here again, a fee-for-service approach is used where fixed charges are applicable for each service. Interested readers will find more information on the concept and design in the project report of the Solar Transitions project (Muchuku et al. 2014).

3.2.3 Solar Multi Utilities for Productive Activities

Another model, in some ways similar to the charging station but larger in capacity is the Solar Multi Utility (SMU) piloted by organizations such as TERI and Mlinda Foundation in over 10 locations in India. The concept is centred on the development of income generating activities in rural locations, with support from electricity produced by the Solar Multi Utility Centre. The SMU is essentially a centrally located solar power plant (without distribution) which is used to power a number of activities such as grinding, juice production, bottling, sealing and packaging, weighing, water purification among others. Individual villagers, farmers associations or women's self-help groups (SHG) use this facility to convert their raw material into an end product which is then sold in the market.

Here again, the charges for each type of service may be defined either in terms of kWh or on a service charge basis. When communicated in kWh, however, customers will compare the cost of electricity to that of the nearby grid and therefore not consider the SMU cost-effective owing to the higher cost of solar-based electricity and the subsidized cost of grid electricity in rural locations. Therefore, a calculation of the cost of each service could be arrived at on a service basis such as the cost of grinding one kilogram of wheat or the cost of bottling hundred bottles of juice.

4   Case Studies

This section aims to showcase some examples of business and delivery models based on the types of systems discussed in the previous section. The objective here is to demonstrate some of the ways in which project implementers have innovated and deviated from standardized modes of electricity delivery to bring value to end consumers. While models where subsidies are provided or direct sales are involved are also part of the broader discussion on off-grid energy service delivery, the focus of this section is limited to discussing the technical and economic aspects at the project level, rather than the methods by which finance is made available for project implementation. Subsidies will of course enhance the viability of the project even more if available.

BOX 1: Fee-for-Service model

The fee-for-service model provides a certain service, such as lighting or mobile phone charging to users for a daily, weekly or monthly fee. The hardware or power plant is owned by a public or private company, or if the model is donor driven, by the implementing organization or local entrepreneur. The operation, maintenance and repair of the system are therefore the responsibility of the owner, who recovers their investment through the fee charged for a service. In some cases, the owner of the system is also known as an ESCO or Energy Service Company. This model is of particular relevance to this chapter because often in its delivery, the fee is a flat charge, delinked from the kWh charge of electricity.

The fee-for-service model has the inherent advantage that the end-users do not need to pay for financing the system and is therefore well suited to low income consumers. Additionally, as operation of the system is the responsibility of a company/operator who is trained and has bulk access to spares (as opposed to household users), repair and maintenance are standardized and faults may be resolved quickly. Finally, in remote locations where it is difficult to find financing for individual systems from banks and other institutions; such a model builds in scale into the system leading to better chances of securing finance.[9]

On the other hand, regular revenue collection remains a challenge and as stated earlier, the lack of ownership among the users may lead to some misuse and damage. However, as demonstrated through the case studies, a careful design of programme operations can assist in reducing some of these risks.

---

[9]While a single charging station may not qualify for a bank loan, the ESCO or private company could aggregate a number of such systems and qualify for a loan.

Four cases from India are discussed here;

- Lighting a Billion Lives (LaBL) Programme (Solar Charging Stations) by TERI, India,
- Solar DC micro-grid initiative under TERI's Norwegian Framework Agreement with Ministry of Foreign Affairs, Norway
- Solar DC micro-grids in Uttar Pradesh by Mera Gao Power, India
- AC pico-grid model of Mlinda Foundation, India

4.1 TERI's Lighting a Billion Lives (LaBL) Programme

The LaBL programme was initiated by The Energy and Resources Institute (TERI) in 2008. The programme aims to disseminate solar lighting devices using a variety of business models to rural consumers and displace the use of kerosene or paraffin lanterns (Krithika and Palit 2013).

The programme operates on a fee-for-service model wherein solar charging stations are installed in a village and operated by an entrepreneur who then rents out the lanterns to customers on a daily fee basis. The rent collected serves as an income for the entrepreneur as well as a fund for operation and maintenance and battery replacement. TERI has designed the charging station, implements the system with its technology partners and then provides training and capacity building support to the entrepreneurs. With over 2,500 villages covered across India and countries in Africa, the LaBL programme is a good example of a scaled imitative in off-grid lighting.

In the case of solar lanterns, the capital cost of solar charging station including one battery replacement is INR 1,30,000[10] (USD 2167) for a charging station with 50 lanterns. The annual maintenance cost per charging station (with 50 lanterns) is INR 5,000 (USD 83), while the practical life of the solar charging station is 3–4 years. If we add the capital cost and maintenance cost (discounted at a 10 % annually), the total cost for a 3-year period is INR 1,42,434 (USD 2374).

We can calculate the kWh generated by the lantern by considering the capacity of the battery inside the lantern, which is a 6 V, 4.5 Ah lead acid battery amounting to 27 Wh. For such batteries at depth of discharge of 60 % is recommended under LaBL programme. Thus the resulting battery capacity is $27 \times 0.6 = 16.2$ Wh. These lanterns usually operate 30 days a month and 12 months a year. Thus for 3 years and 50 lanterns, the total kWh = $(16.2 \times 50 \times 30 \times 12 \times 3)/1000 = 875$ kWh.

On dividing the total kWh delivered by the total cost per charging station calculated earlier, we get the cost of generation as 162.7 INR/kWh (2.71 USD/kWh).

---

[10]All costs are for the base year 2013–2014.

4.2 DC Microgrids Under the Norwegian Framework Agreement

As a second case, we consider the DC micro-grids implemented under the Norwegian Framework Agreement by TERI in a few districts in Uttar Pradesh. A DC-microgrid is operated by a village level entrepreneur and in the case of this project, the entrepreneur also invested 45 % of the total cost of the system while the remainder was supported by TERI. Typically, each DC microgrid connection provides a total of 3 W of LED lights for a period of 5–6 h per day, which is provided in the evening to night hours. This may comprise of a single 3 W LED or two LEDs of 1 W and 2 W each, depending on the needs of the user. The users typically pay INR 5 per (USD 0.83) day or INR 150 (USD 2.5) per month for the service.

In most villages where such systems have been installed, the grid is usually unavailable during the evening and many customers of the microgrid connection are shopkeepers, petty traders and handloom workers in the market area of the village. A few extra hours of light can yield significantly higher revenues for the business and therefore a price of INR 5 (USD 0.83) per day is acceptable. The cost of service on the other hand is INR 5 (USD 0.83) per day for $3 \times 5 = 15$ Wh of energy. This works out to about 333 INR/kWh (5.55 USD/kWh), significantly higher than the grid tariff. However, it is exactly this difference in tariff which has enabled the DC microgrid entrepreneur to get a complete return on her investment in a time span of about one year.

4.3 Mera Gaon Power (MGP): DC Microgrids

MGP is a private company and has been working in Sitapur district of Uttar Pradesh for the last 3 years and has over time developed a DC microgrid model to provide basic electricity access to rural people for lighting and mobile phone charging. The model is low-cost (around INR 55,000 (USD 915) for 25–30 connections) and each connection includes two 1 Watt LED lights and one mobile charging point (total 4 W of load). Users may choose to pay extra for additional lights, however, a maximum of eight 1 Watt LED is provided per household. One house per grid is chosen as the System House where the panels are installed and batteries and charge controllers are kept in a secure wooden box with a lock arrangement.

MGPs focus is on strengthening operations and ensuring timely collection of tariff in a manner that is favourable to both MGP and its customers. A Joint Liability Group (JLG) model, often used by Micro Finance Institutions (MFIs) has been developed, with all the users of a single micro-grid acting as one JLG. Such an arrangement has reportedly led to collection efficiencies of over 90 %.

Each user pays MGP the user fee and if any user is absent or unable to pay, the JLG has to pay on that user's behalf. Non-payment leads to immediate disconnection. The connection charge is INR 50 (USD 0.83) and tariff collections of

INR 25 (USD 0.42) per week are made on a prepaid basis. Users may also choose to pay in advance for more than one week as well. In case of disconnected systems (due to non-payment), a reconnection charge of INR 40 (USD 0.67) is levied.

The highlight of the MGP model besides the operational efficiency brought in by the JLG mechanism is the manner in which the tariff collection process has converged with the existing modes of payment of lighting services. A weekly payment as opposed to a larger monthly or bimonthly bill leads to transactions which are easier for rural consumers to manage with their limited and fluctuating incomes. This is possibly one of the reasons behind the low collection efficiency by electric utilities from grid connected consumers who receive larger monthly or bimonthly bills. A lesson from this model could be utilized even in the grid-connected areas.

The second highlight is with respect to the tariff itself, which like in the case of TERI's DC micro-grids is designed to match the coping cost of the user rather than the grid-tariff (which is cross-subsidized and does not factor in environmental costs). A weekly charge of INR 25 (USD 0.42) is levied for a basic connection of 4 W (available for 7 h), amounting to 90 INR/kWh (1.5 USD/kWh), again much higher than the lifeline tariff of INR 2–3/kWh for grid electricity. Once again we find that this cost of INR 25 (USD 0.42) per week is equal to, or lower than the coping cost of purchasing kerosene (INR 100–150 per month for a household) or other alternatives and is, therefore, beneficial to the user as well as supports the operations of MGP as a for-profit company with reported paybacks of between 2 and 3 years per system. It is also important to note that for the same or lower cost, the users are getting better quality light than they would get from a kerosene lamp and this also is an important decision-making criteria in case the costs of solar lighting and kerosene are the same.

4.4   Mlinda Foundation's AC Pico-Grids

Mlinda Foundation's Solar Pico Grids programme covers households, schools, markets and productive power segments. In comparison to many rural locations in India, the paying capacities and future income generating opportunities of the residents of these villages are relatively better, enabling the creation of new types of business models which are less dependent on subsidies.

In partnership with National Bank for Agriculture and Rural Development (NABARD), Mlinda has made it possible for people to avail loans to buy the solar installations. Loans are available to a group of households collectively organized into Joint Liability Groups (JLGs). The end-user repays through affordable installments over a period of four years from the direct savings accrued from non-usage of kerosene for lighting (Schäfer and Kammen 2014). The grids are owned by the JLG after repaying the banks. Here again, the coping cost is the cost of procuring kerosene, in the range of INR 100–150 per household per month. It is exactly this amount of money which repays the loan for a solar mini-grid, making the transaction feasible from the point of the view of the users. The cost of generation from such systems has been estimated at around 120 INR/kWh (2 USD/kWh), once again demonstrating that while the kWh cost is high, the actual payments made by users is within their budget and coping costs.

These JLGs are linked to NABARD and the repayment is done by the group through monthly installments. The entire group is accountable for repayment of the loans, which reduces the chances of delayed payments and bad debts (TERI, 2014).

It is critical to note here the methods by which the end-users have the opportunity to own their power generating system, i.e. through a combination of loans and subsidy. This is important because individuals in rural areas who are engaged primarily in informal occupations often do not have the necessary credit history to avail loans from banks. The JLG model is an attempt to enhance the bankability of these individuals through collective applications for loans and repayment. More significantly, owing to the small quantum of each installment, rural customers are able to afford the payment as opposed to a large upfront investment. From the point of view of the financing institution, such an arrangement reduces the risk for the bank by ensuring collective guarantees for repayments as well as scale. In summary, this is another example which showcases how the tariff in kWh terms has not been considered as a benchmark to provide an energy service. Rather, the same cost has been converted into monthly installments which can lead to eventual ownership of the system.

BOX 2: Pay-as-you-go models

Other models which are innovative in terms of end-user financing and do not involve the kWh-based pricing structure are pay-as-you-go models. In such models the customers are required to make a small down payment upfront, typically 10–30 % of the fully financed cost, to receive the solar product or to have it installed at their premises. Customers are further required to prepay to use the solar product through a mobile-based energy credit model. Energy service is denied to the customer by the technology embedded within the product if the customer's prepaid balance has been used or if it expires, enabling access again when the customer adds prepaid credit to their account. In some models, the customer pays off faster and ultimately owns the energy asset through a rent-to-own option or through leases (Winiecki and Kumar 2014).

For instance, Simpa networks (India) sells distributed energy solutions on a "Progressive Purchase". The customers make a small initial down payment for a high quality solar PV system which can power four lights and provide mobile charging. They then pre-pay for the energy service, topping up their systems in small user-defined increments using a mobile phone. Each payment for energy adds towards the final purchase price. Once fully paid, the system unlocks permanently and produces energy. This model is operational in Karnataka (Simpa Networks 2014).

The business model of Simpa primarily runs through appointed agents to whom it sells prepaid energy credits. A microenterprise can register as a Simpa energy credit agent which acts as the interface between Simpa and the customer. The agent registers his name, address, telephone number, mobile number in Simpa's software. He buys energy credits from Simpa via local bank deposit, while the end customers pay cash to the agent. Agent sends an SMS to Simpa with customer's unique ID, agent's ID and payment details. Simpa's software validates the transaction based on the customer and agent's registered data and sends a unique, single use numeric code back to the end customer via SMS. Customer enters the usage code into his or her own solar system via the embedded user interface. Solar device unlocks use of the system for the prepaid amount of time (Winiecki and Kumar 2014).

5 Conclusion

This chapter has attempted to showcase two distinct methods by which developers are designing their business models for off-grid solar PV systems. While the more traditional method involves the calculation of a cost of generation, it has been found that in remote and rural areas where off-grid systems are implemented, service-based approaches are becoming increasingly popular. We have discussed how service-based charges are in some cases more cost-effective, are easier to communicate to rural consumers and also mimic their existing expenditure patterns, which could lead to higher collection efficiencies. Linked to this, a second important focus in this chapter has been on the value of incorporating socio-economic considerations in designing business models. While tariff is one aspect, ownership, maintenance, spread of households, availability of finance and so on are also being structured in innovative ways that are better suited to end-users in rural locations as demonstrated through the case studies.

While we have covered a few case studies as examples, there are a much larger number of such innovative models and interested readers may refer to the Energy Map operated by Santa Clara University (http://energymap-scu.org/) and Global Network on Energy for Sustainable Development Energy Access Knowledge Base (http://www.energy-access.gnesd.org/) for a comprehensive listing of such off-grid energy projects.

We expect that the key take-away from this chapter is the importance of thinking about developing business models which are at times significantly different from the utility model of conventional electricity supply. The conventional grid-based electricity supply may be able to provide power at a lower cost than an off-grid power plant, but this is as a result of a large number of supporting factors such as large consumer bases, cross-subsidies through regulation and other forms of policy and financial support from the country governments. The off-grid system developer on the other hand, may not be receiving these incentives, but is providing a service in regions where the coping cost of non-availability of the grid can be quite high.

From the end-users point of view therefore, it is important to evaluate not just the relative cost of grid supply, but the total coping cost of securing modern energy services. We have attempted to demonstrate how a business model is developed not just based on the cost of generation of power but on a range of other factors which create an alternative definition of 'grid parity' for the rural consumer. The point of convergence of the user's needs and capacity to pay with the developer's requirements for a healthy return in off-grid projects can only be achieved by thinking beyond standard metrics of measuring electricity delivery.

Acknowledgements The authors gratefully acknowledge the funding support received to undertake the study as part of the research project titled "Decentralized off-grid electricity generation in developing countries: Business models for off-grid electricity supply", funded by the Engineering and Physical Sciences Research Council/Department for International Development (research grant EP/G063826/1) from the Research Council United Kingdom Energy Programme. The authors would like to thank the entire team of professionals from TERI and our partner organizations in the off-grid energy sector who have worked with us over the years to build innovative off-grid energy systems and share the experiences from the field. We would like to thank Ms. Sangeeta Malhotra, Project Associate in TERI, for her assistance with illustrations and editing. Authors of the reference materials are also gratefully acknowledged.

References

Bhattacharyya, S. C. (2014). Business issues for mini-grid-based electrification in developing countries. *Mini-Grids for Rural Electrification of Developing Countries* (pp. 145–164). Heidelberg: Springer International Publishing.

Central Electricity Authority. (2011). Progress report of village electrification. Retrieved October 18, 2014 from http://www.cea.nic.in/reports/monthly/dpd_div_rep/village_electrification.pdf. .

Central Electricity Regulatory Commission. (2014). Depreciation Schedule. Retrieved October 14, 2014 from http://www.cercind.gov.in/131205/appendix_2.pdf.

Chaurey, A., Krithika, P. R., Palit, D., Rakesh, S., & Sovacool, B. K. (2012). New partnerships and business models for facilitating energy access. *Energy Policy, 47*, 48–55.

Crundwell, F. (2008). Finance for engineers: Evaluation and funding of capital projects. Berlin: Springer.

ESMAP. (2001). *Best practice manual: Promoting decentralized electrification investment.* Washington, DC: The World Bank.

IED. (2013). *Identifying the gap and building the evidence base on low carbon mini-grids: Support study on Green Mini-grid development.* Francheville: Innovation Energie Developpement.

Kandpal, T. C., Garg, H. P. (2003). Financial evaluation of renewable energy technologies. New Delhi: MacMillam India Limited.

Krithika, P. R., Palit, D. (2013). Participatory Business models for off-grid electrification. *Rural Electrification through Decentralised Off-grid Systems in Developing Countries* (pp. 187–225). London: Springer.

Ministry of New & Renewable Energy. (2013). Amendment in the bench mark cost for "Off-grid and Decentralized Solar Applications Programme" being implemented under the Jawaharlal Nehru National Solar Mission (JNNSM) during 2013–14. Retrieved October17, 2014 from http://mnre.gov.in/file-manager/UserFiles/amendmends-benchmarkcost-aa-jnnsm-2013-14.pdf.

Ministry of Power. Retrieved October 20, 2014. http://powermin.nic.in/rural_electrification/definition_village_electrification.htm.

Muchuku, C., Ulsrud, K., Winther, T., Palit, D., Kirubi, G., Saini, A., et al. (2014). *The Solar Energy Centre: An Approach to Village Scale Power Supply*. Norway: University of Oslo.

OECD/IEA. (2014). Technology Roadmap: Solar Photovoltaic Energy. Paris: International Energy Agency.

Palit, D. (2013). Solar energy programs for rural electrification: Experiences & lessons from South Asia. *Energy for Sustainable Development, 17*(3), 270–279.

Palit, D., Malhotra, R., & Kumar, A. (2011). Sustainable model for financial viability of decentralized biomass gasifier based power projects. *Energy Policy, 39*(9), 4893–4901.

Palit, D., & Singh, J. (2011). Lighting a billion lives—Empowering the rural poor. *Boiling point, 59*, 42–45.

Rabl, A., Fusaro, P. (2001). Economic and financial aspects of distributed generation. Distributed Generation: The Power Paradigm for the New Millennium 203.

Schäfer, M., & Kammen, D. (2014). *Innovating energy access for remote areas: Discovering untapped resoures*. Berkeley: University of California.

Sharma, K. R., & Palit, D. (2014). Decentralising the solar lighting provision: A case study of a solar lantern delivery model from Kenya. *Boiling Point, 63*, 2–5.

Short, W., Packey, D. J., Holt, T. (2005). A manual for the economic evaluation of energy efficiency and renewable energy technologies. Honolulu: University Press of the Pacific.

Simpa Networks. (2014). Retrieved October 21, 2014 from http://simpanetworks.com/.

Tenenbaum, B., Greacen, C., Siyambalapitiya, T., & Knuckles, J. (2014). *From the bottom up: How small power producers and mini-grids can deliver electrification and renewable energy in Africa. Directions in development*. Washington, DC: World Bank.

Thakur, N., Sharma, A., Mohanty, P., Sharma, K. R., & Parmar, P. (2011). Solar lighting systems in India: Types, applications and performance assessment. *Journal of the Solar Energy Society of India, 21*, 64–86.

Uttar Pradesh Electricity Regulatory Commission. (2013). Suo Motu Determination of Annual Revenue Requirement (ARR) And Tariff For FY 2013–14 for Paschimanchal Vidyut Vitran Nigam Limited. Retrieved October 20, 2014 from http://www.pvvnl.org/RTI/TARIFF/Tariff_2013-14/PVVNL%20Suo-motu%20ARR%202013-14_31052013.pdf. .

Ulsrud, K., Winther, T., Palit, D., Rohracher, H., & Sandgren, J. (2011). The solar transitions research on solar mini-grids in India: Learning from local cases of innovative socio-technical systems. *Energy for Sustainable Development, 15*, 293–303.

Winiecki, J., & Kumar, K. (2014). *Access to energy via digital finance: Overview of models and prospects for innovation*. Washington, DC: Consultative Group to Assist the Poor.

Hybrid Energy System for Rural Electrification in Sri Lanka: Design Study

Iromi Ranaweera, Mohan Lal Kolhe and Bernard Gunawardana

Abstract Off-grid hybrid renewable energy based power system for rural electrification has become an attractive solution for areas where grid electricity is not feasible. Depending on the availability of the resources, single energy source or a combination of several sources, including an energy storage system is used in an off-grid hybrid energy system. As there can be several candidates, the optimum sizing of the components against resource potential, cost and reliability is a vital issue. This chapter presents a study of optimum sizing of a renewable sources based off-grid hybrid energy system that is designed for electrifying a rural community in the Siyambalanduwa area in Sri Lanka. The community consists of approximately 150 households with a daily electricity demand about 270 kWh and night time peak of 25 kW. This region receives abundant of solar irradiation with an average of 5.0 $kWh/m^2/day$. The annual average wind speed of this region is 6.3 ms^{-1} which results in wind power density of 300 W/m^2 at a height of 50 m above the ground. The total net present cost of a configuration is calculated for 20 years of system's lifetime to examine the lowest energy cost option. It is found that the combination of wind turbines, PV system, a battery bank and a diesel generator made the optimum hybrid system having capacities wind—40 kW, PV—30 kW, battery bank —222 kWh and the diesel generator—25 kW. This system can supply electricity at an approximate levelized cost of 0.3 $/kWh. It can supply the demand without change in energy cost more than 0.1 $/kWh even though the annual average wind speed varies in the range of 4.5–6.3 ms^{-1}. Consequently, influence on energy cost for changes in annual average solar irradiation in the range of 4.0–5.5 $kWh/m^2/day$ has been found to be negligible. The energy cost analysis of the project has also been made considering off grid operation of hybrid systems for first 10 years and grid connected operation for next 10 years. It is found that the hybrid system is economically viable whether it is operated in off-grid or grid connected mode.

I. Ranaweera · M.L. Kolhe (✉) · B. Gunawardana
Faculty of Engineering and Science, University of Agder, PO Box 422,
Kristiansand NO 4604, Agder, Norway
e-mail: mohan.l.kolhe@uia.no

© Springer International Publishing Switzerland 2016
P. Mohanty et al. (eds.), *Solar Photovoltaic System Applications*,
Green Energy and Technology, DOI 10.1007/978-3-319-14663-8_7

165

1 Introduction

The electricity sector in Sri Lanka is primarily based on hydro power and imported fossil fuels. Currently hydro energy contributes 40.5 % of the total installed capacity while the contribution from the thermal energy is 49 %. The remaining 10.5 % is from the renewable energy sources such as mini hydro, bio-energy, wind energy and solar energy. The electrification rate in Sri Lanka by the end of 2013 was 96 % and rapid progress in this sector has been observed over last few years towards achieving a 100 % rate (Annual performance report 2013). Mini hydro plants play a vital role in electrifying rural communities where the main grid is not accessible. A large number of off-grid village hydro schemes with capacities ranging from 3–50 kW are in operation benefiting 17–116 customers (Public Utilities Commission of Sri Lanka 2012). However, hydro resource is not available in everywhere. Photovoltaic systems have gained attention in such regions. Because of the intermittent nature of the solar resource, isolated solar home systems (SHS) require energy storage. The capability of these systems in supplying the load is limited to a few lighting loads and other loads such as black and white television and an iron. The quality of the supply and the reliability can be improved by interconnecting several energy generating systems known as a hybrid system. A hybrid system can form a micro-grid, which can be easily connected to the main grid later if required. However, proper designing of a hybrid system is challenging as it can contain variety of technologies. Generally, it requires calculation of the total life cycle cost of different alternatives and the design configuration that can supply the given load with required level of availability at a lowest cost. The alternatives can consist of different mixes of generation sources with different capacities. The robustness of the system and the changes in the system variables such as load, energy resource potential need to be accounted for, when finding the optimum design.

The focus of this chapter is to investigate the techno-economical optimum design of a hybrid system for electrifying a given rural community in Sri Lanka. The levelized energy cost is calculated and compared with the grid electricity price. The economic feasibility of implementing such project is discussed. Finally, analysis is made assuming the micro-grid will be connected to the grid after few years of isolated operation to investigate the impacts on return on investment.

2 Load Demand of a Rural Community

There are several rural areas in Sri Lanka, which are not yet electrified with abundant amount of solar energy, hydro energy, and/or wind energy. Even though SHS, small wind turbines and micro-grids powered by either micro-or pico- hydro, dendro, thermal plants or bio-gas power plants are in operation in some of the rural regions, attention on hybrid energy systems is insignificant (Public Utilities Commission of Sri Lanka 2013). In Sri Lanka, Ceylon Electricity Board (CEB) is the authorized body for supplying electricity across the country. Recently, CEB has announced a list of villages that will not get grid access in the near future. The Ministry of Power and Energy has instructed the Sustainable Energy Authority to take necessary measures to provide off-grid renewable energy solutions to those households (Public Utilities Commission of Sri Lanka 2013). In this list, there are 1072 villages, consisting of 37800 households in different parts of the country. In this study, we have considered Monaragala district in Uva province for investigating the techno-economics of a hybrid energy system. This region covers 258 villages that will not get grid electricity in the near future. It is one of the economically poorest regions in Sri Lanka with a population of approximately 0.5 million whose income is derived mainly from agricultural activities (Population of Sri Lanka by district 2012). There are many villages in this district without electricity, which are located far from the urban centres, such as some villages of the Siyabalanduwa Divisional Secretariat (DS). It consists of 48 villages with a total population of 55300, with 13600 households. Among these households, 5750 households do not have access to the electricity (Department of Census and Statistics 2012). We selected one of the villages from the Siyabalanduwa DS, which is located at 6.76°N latitude and 81.54°E longitude for this study.

2.1 Electric Load

A village comprising approximately 150 family houses is considered and the hypothetical load profile of the village is derived based on the following assumptions and referring the load profiles available in the literature for different rural electrification projects implemented in developing countries (Nayar 2010; Intelligent Energy-Europe 2008; Kirubi and Jacobson 2009).

- It is supposed that the village comprises 10 rich families, 50 medium income families and 90 low-income families. This assumption is made because majority of the people living in Siyambalanduwa are economically poor.
- The village consists of a community centre, temple, pre-school, primary school, 2 shops, street lights and two rice mills which consume daily electrical energy of about 60 kWh.
- Wealthy families use electricity for operating energy efficiency light bulbs, colour television, cassette, DVD player, fans, refrigerator, water pump, computer and an iron. Usually electricity is not used for cooking in rural villages of Sri Lanka. Instead, they use firewood, because it is widely available at zero cost. The daily energy consumption of this type is assumed as 4 kWh.
- Medium income families use electricity for operating energy efficiency bulbs, radio, television and an iron. A medium income family's daily energy consumption is assumed as 2 kWh.
- The daily energy consumption of low-income families would be very low. They will use electricity for fulfilling the basic requirements such as lighting,

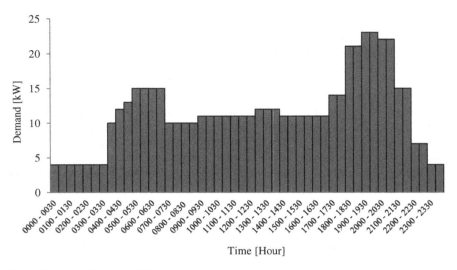

Fig. 1 Load profile of the village

communication (radio and television) and maybe for ironing clothes. Families in this category have very small houses; therefore their lighting load will also be very small. Thus, the daily energy consumption for low-income families is assumed as 0.8 kWh.

Since the village does not contain many commercial industries that create a larger demand during the daytime, the load profile of the village will have a low early morning peak and a high night peak with the flat daytime load, which is lower than the early morning peak. The load curve derived for this village based on the above assumptions is given in Fig. 1. The maximum demand around 25 kW and daily energy consumption around 270 kWh is observed in the profile. The monthly mean temperature of this area ranges from 25 C to 27 °C throughout the year. Thus, this area is not affected by seasonal variations. In addition, the day length in Sri Lanka does not vary significantly throughout the year due to its geographical location near the equator. Therefore, a constant load profile can be assumed over the year.

3 Energy Technologies for Hybrid Energy System

3.1 Solar Energy

Sri Lanka is an island located nearer to the equator; therefore, it receives plentiful solar irradiation throughout the year. The monthly averages of the daily irradiation in this region obtained from the NASA Surface Meteorology and Solar Energy

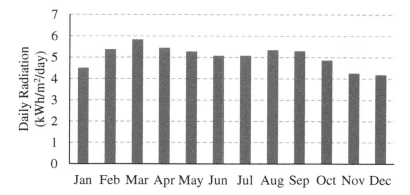

Fig. 2 Monthly averaged insolation incident on a horizontal surface and clearness index

database are shown in Fig. 2. According to this data, the area receives annual average of daily solar irradiation about 5.0 kWh/m²/day. From the figure, it can be seen that there is no significant change in the monthly averages over a year.

3.2 Wind Energy

Sri Lanka is an island with substantial wind energy resources. Its wind climate is primarily determined by two Asian monsoons, the southwest and the northeast monsoons. The southwest monsoon continues from May till early October while the northeast monsoon continues from December to February. The monthly averages of wind speed at 50 m anemometer height obtained from the NASA Surface Meteorology and Solar Energy database is shown in Fig. 3. The annual average wind speed of this region is 6.3 ms⁻¹. The RETScreen database gives the wind speed data at the same location, but for an anemometer height of 10 m. These wind speeds at different height levels are used to calculate the surface roughness of the site.

3.3 Battery Energy Storage

An off-grid hybrid system requires storage to store the excess energy from the renewable sources for later utilization when enough power is not produced by the renewable sources. Batteries are the most common storage medium used in renewable applications. Several types of batteries are available such as lead-acid, nickel cadmium, lithium, zinc bromide, zinc chloride, sodium sulphur, nickel hydrogen and vanadium redox flow batteries (Zaghib et al. 2015; Cho et al. 2015;

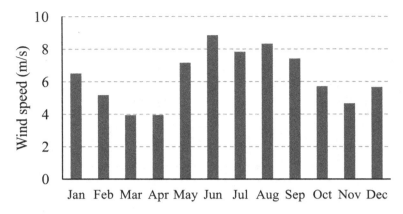

Fig. 3 Monthly averaged wind speed

Skyllas-Kazacos and McCann 2015). Among these, lead-acid batteries are widely used in off-grid systems due to their low cost, high voltage per cell and good capacity life. The life of a battery is primarily affected by the depth of discharge and the operating temperature. Depth of discharge is the level at which batteries are discharged in a cycle before they are charged again. Overcharging of the battery results in electrolysis of the water and overheating while deep discharging accelerates the battery degradation. Hence, the battery state of charge (SOC) should be maintained within the safe limits to maximize the battery lifetime (Farmann et al. 2015). The battery capacity is the amount of energy that can be withdrawn from starting to fully charged state. It depends on the rate at which energy is withdrawn from the battery. The higher the discharge current, the lower the capacity. When batteries are left standing without charging, the batteries lose charge slowly by self-discharging. This occurs due to the reactions within the cells of a battery. The self-discharge rate depends on the temperature, the type of battery and their age. As batteries get older, self-discharge rate increases. Batteries require regular maintenance in order to maintain proper operation during their lifetime. The maintenance requirements for batteries vary significantly depending on the battery design and application. Generally batteries require maintenances, cleaning of cases, cable and terminals, tightening terminals, distilled water addition and performance checks. Performance checks include specific gravity recordings, conductance readings, temperature measurements, cell voltage reading and capacity test.

3.4 Diesel Generator

Diesel generators are important in off-grid hybrid systems to improve the availability of the electricity supply at a low cost. As diesel generators are despatchable, they can be used to supply the load when the energy productions from the renewable sources are low or the state of the charge of the battery bank is not sufficient for supplying the load. Having a diesel generator in the system can reduce the size of the storage, which is relatively expensive. Thus, a reduction in cost can be expected by adding a diesel generator, which will be operated only for a limited number of hours over a year. Usually diesel generators operate at higher efficiency near full load. Thus, it is recommended to operate the generator above a certain load factor for maintaining the proper efficiency of the energy conversion hence lowering the fuel cost by reducing the utilization of fuels. Nevertheless, the despatch strategy of the generator in a hybrid system depends on several factors such as size of the generator and battery bank, the fuel price, operation and maintenance cost of the generator and the amount of renewable power available in system. It can either be operated at full load when required and transfer excess energy to the battery bank, or at the required load level that cannot be supplied from other sources. The first strategy is known as cycle discharging and the other is known as load following (Mohammed et al. 2015). When optimizing the system size, despatch strategy of the generator also needs to be taken into consideration as it greatly affect the total cost.

4 Hybrid System Configuration

Different technologies have different operating characteristics. Hence, power electronic converters are used with these systems to provide the interface to the AC grid. There are several ways of connecting these components to form the grid. They can be classified mainly into three categories: DC coupled, AC coupled, and hybrid coupled (Nehrir et al. 2011). The AC coupled configuration shown in Fig. 4 is considered in this work. In this configuration the candidate technologies are coupled to the AC bus through power electronic converters.

5 System Cost

In order to examine the best configuration that can provide the electricity to the community at a lowest cost with the required level of availability, the lifetime cost of the different alternatives needs to be calculated and compared. The system should be sized such that sufficient energy is available to supply the demand and the cost of the energy produced from the system should be minimized as much as possible. The life cycle cost calculation entails all the costs incur over the lifetime of the project that includes installation capital, operation and maintenance cost, replacement cost and all other costs. The summary of the estimated capital cost, operation and maintenance cost (O&M), and the replacement cost of all the system components is given in Table 1. The cost estimation is done based on the current prices of the components and the services.

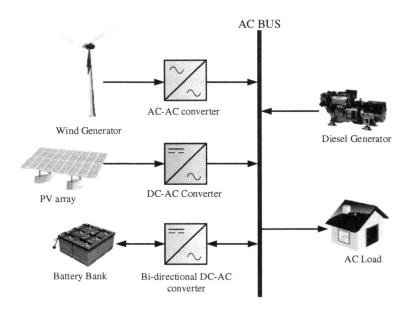

Fig. 4 AC coupled hybrid system configuration

Table 1 Component costs

Component	Capacity	Capital cost [USD/unit]	O&M cost	Replacement cost [USD/unit]
PV system	10 kW	29000	10 $/year	23000
Wind turbine	10 kW	25000	500 $/year	20000
Battery	1156 Ah	1400	10 $/year	1400
Generator	10 kW	6300	0.5 $/hr	6300
	15 kW	9700	0.6 $/hr	9700
	21 kW	12000	0.7 $/hr	12000
	25 kW	14000	0.8 $/hr	14000
	29 kW	14500	0.9 $/hr	14500
Solar inverter	10 kW	3500	0	3500
Wind energy inverter	10 kW	4000	0	4000
Battery inverter	4.2 kW	3400	0	3400
	5 kW	4000	0	4000

6 Technical Parameters of the System

The technical parameters of the system components used in the study are given in Table 2.

Table 2 Technical details of the components

PV system	
Model	Canadian solar CS6P- 240P
Peak power	240 W
Derating factor	80 %
Slope	6.81°
Azimuth	0°
Ground reflectance	20 %
Temperature coefficient	−0.43 %/°C
Nominal operating temperature	45 °C
Efficiency at standard test conditions	14.92 %
Lifetime	20 years
Wind turbine	
Model	Hummer H8.0–10 kW
Rated power	10 kW
Hub height	15 m
Lifetime	20 years
Battery	
Nominal voltage	6 V
Nominal capacity	1156 Ah
Lifetime throughput	9645 kWh
Round trip efficiency	80 %
Min. state of charge	40 %
Float life	15 years
Maximum charge rate	1 A/Ah
Max. charge current	41 A
Batteries per string	8 (48 V DC bus)
Diesel generator	
Lifetime (operating hours)	25000 h
Minimum load ratio	50 %
Fuel	Diesel
Fuel cost	0.96 USD/l
Converter	
Lifetime	20 years
Efficiency	94 %

7 Optimum System Design

After we obtain the relevant information about the potential of the resources, the technical details of the components and all relevant costs for building a hybrid system, the net present value (NPV) of the total lifetime cost of the system is calculated for different alternatives. In order to find the optimum design the commercial software HOMER is used. HOMER is a computer model that simplifies the task of evaluating design options for both off-grid and grid connected power systems for remote, stand-alone and distributed generation applications. It facilitates a range of renewable energy and conventional technologies, including solar PV, wind turbine, hydropower, generator (diesel, gasoline, bio-gas), battery bank and hydrogen. HOMER's optimization analysis algorithms allow the user to evaluate the economic and technical feasibility of a large number of technologies . The sensitivity analysis can also be performed in HOMER and it allows finding the effect of uncertainty in the input variables to the energy cost and the optimal configuration.

The hybrid system shown in 4 is modelled in HOMER and the details given in Tables 1 and 2 are given as inputs to the model. Then optimization calculation is carried out for several alternatives with different combinations of technologies and different capacities. It calculates the total net present cost of all the feasible system configurations. The system having the lowest total NPV is the optimum design. When doing an optimization, it is important to find a system which is robust, so that the variations in the input variables do not affect the system costs and the performance. This is called sensitivity analysis. The sensitivity analysis is carried out in HOMER when selecting the optimum configuration. The sensitivities for variation in following parameters are investigated.

- Annual average solar irradiation and wind speed in the range of 4.5–5.0 kWh/m2/day and 5.5–6.4 m/s.
- Capital cost of PV and Wind system in the range of ±15 %.

The system configuration given in Table 3 is found as the optimum system based on HOMER calculations. The system consists of PV array, wind turbines, battery bank and a diesel generator.

Table 3 Optimum hybrid configuration	PV capacity	30 kW
	Wind turbines- 10 kW	4
	Diesel generator capacity	25 kW
	Battery bank	222 kWh
	Converter capacity	25

8 System Performance

According to the HOMER results, this project requires an initial capital of approximately $296000 and the NPV of the total life time cost (TLC) of the project is $553000. This system can supply the energy for a levelized cost of 0.34 $/kWh. The levelized cost of energy (LCOE) is the cost that, if assigned to every unit of energy produced (or saved) by the system over the analysis period, will equal to the TLC when discounted back to the base year. Figure 5 illustrates how the costs are distributed among different components of the system.

Monthly average electric power production from each of the system components in the hybrid system is shown in the Fig. 6. It can be seen that the wind turbines generate the largest percentage of the power. Especially the power generated from the wind turbines is considerably higher during the months from May to September. On the contrary, the average power generated from the wind turbines is relatively small during the period of March to April. Therefore, the diesel generator has to

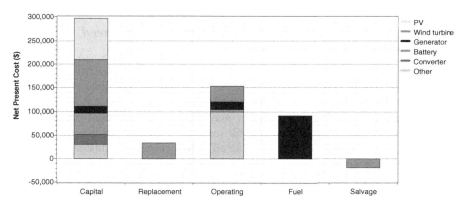

Fig. 5 Cost summary of the project

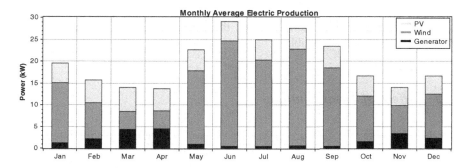

Fig. 6 Monthly electricity production from different system components

produce much more energy during March and April than the other months. As can be seen in the figure, energy generation from the renewable systems is considerably higher during May to September. However, the diesel generator still contributes for supplying the load. That implies the diesel generator is still required to supply the peak demand during this period.

The renewable fraction of this system is 0.91. In this 91 %, wind turbines generate 67 % and the PV system generates 24 %. The rest is generated from the diesel generator. The excess energy percentage of the system is 35 %. The capacity shortage of this system is only 0.1 %, which is very small resulting only 9 h of load shedding during the whole year caused due to inability of producing enough power to meet the load. The generator despatching strategy used here is the load following criteria; therefore, the generator must be operated such that it produces only the required amount of power to cover the shortage capacity that cannot be supplied from the renewable systems or battery bank to meet the load. Further, the capacity factor of the PV system is 15.8 %, which is relatively low and for wind turbines, it is 33.1 % while for the diesel generator it is 7.5 %.

8.1 Effect of Changes in Annual Average Wind Speed and Solar Resource Changes

The selected hybrid system configuration is optimized for the annual average wind speed and the annual average of daily irradiation 6.3 ms^{-1} and 5.0 $kWh/m^2/day$ respectively. Even though this system may not be the optimum configuration outside these nominal values, it can still meet the demand, but then the cost of energy is different from the value obtained before. Figure 7 illustrates how the energy cost of this system varies if annual average wind speed and solar irradiation vary. As shown in the figure when the wind speed and solar irradiation increases, the LCOE decreases. This is due to the increase in the renewable fraction as shown in Fig. 8. As the renewable systems generate more energy, the amount of energy required from the diesel generator decreases. Hence, the fuel cost also decreases and so the LCOE. Change in the LCOE for the wind speed in the range of 4.5–6.5 ms^{-1} is about 0.1 \$/kWh (13 Rs/kWh). However, for the variations in solar irradiation within the range of 4–5.5 $kWh/m^2/day$, the change in cost is only 0.01 \$/kWh (1 Rs/kWh) which is negligible.

8.2 Effect of Load Changes

At the beginning, the load profile is derived based on certain assumptions. Thus, there can be variations in this load profile in comparison with the actual load profile, which can be found only after implementing the project. Further, load growth is

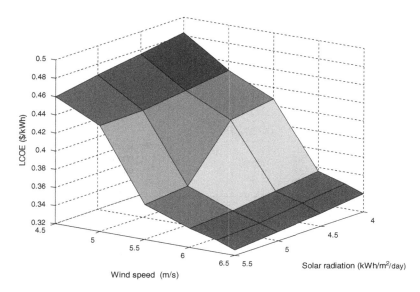

Fig. 7 Effect of changes in annual average wind speed and solar irradiation on LCOE

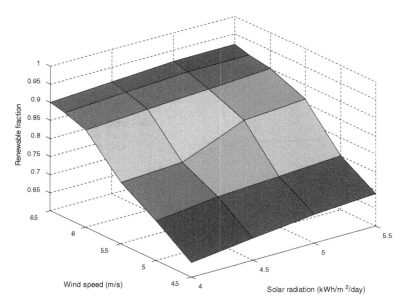

Fig. 8 Effect of changes in annual average wind speed and solar irradiation on renewable fraction

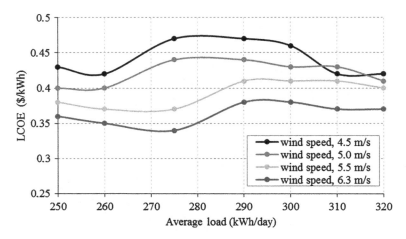

Fig. 9 Variation of LCOE for different daily average load

expected over the time. Thus, analysis is done to observe the variation of the LCOE for variations in the daily load. The results are illustrated in the Figs. 9 and 10.

In Fig. 9, it can be seen that the selected hybrid system can supply the load without significant change in the cost of energy at a certain annual average wind speed even though the daily average load increases to 320 kWh/day. From Fig. 10, decrease in renewable fraction with the increase in demand can be observed. That means the hybrid system meets the growing demand by increasing the power generation by the diesel generator. However, there is no significant change in the

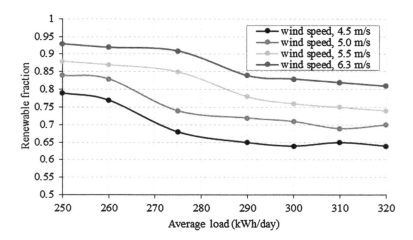

Fig. 10 Variation of renewable fraction for different daily average load

LCOE as load grows for a certain annual average wind speed. Although average annual wind speed drops to 4.5 m/s, the maximum change observed in the LCOE is about 0.15 $/kWh within the selected range in the load growth.

9 Economic Viability

Development of a rural electrification project based on a renewable hybrid energy system in Siyambalanduwa region entails an initial capital investment of approximately $296000. This system can feed approximately 150–175 households, including public utilities and several small businesses. This type of rural electrification scheme can be developed as either private sector based, government based or as an integrated development (Urpelainen 2014). However, the government subsidies will be required to make the service affordable to the end users and to ensure the sustainability of the system even though the project is developed in either way. Because the levelized cost of electricity (0.35 $/kWh) is substantially higher than the average price of the electricity from the national grid which ranges between 0.05–0.35 $/kWh depending on the number of monthly units consumed for domestic user. Nevertheless, the electricity price of rural electrification schemes cannot be compared with the national grid tariff, which already incorporates subsidies, particularly in developing countries.

Table 4 illustrates how the LCOE can be brought down with the capital subsidies from the government or non-governmental organizations (NGOs). Other than capital investment based subsidy, several subsidy schemes are available, for example connection based, output based and operation based (Rolland and Glania 2011). Here we have only considered capital investment based subsidies. As mentioned in Table 4, the LCOE can be reduced by 0.1 $/kWh if subsidy is available for covering 40–50 % of the capital cost.

Solar Home Systems provide huge benefits to the people in rural communities where national grid electricity is not available. These systems are available from 10–60 Wp and the prices vary depending on the size and the complexity of the system (Public Utilities Commission of Sri Lanka 2013). However, the costs of these systems are typically higher and cannot be afforded by low-income rural inhabitants for a onetime payment. Therefore, several loan schemes are available to make these systems viable for rural low-income families in Sri Lanka. The cost of energy generated by these systems typically lies in the range of 0.28–0.30 $/kWh, which is higher than the energy price of the micro-grid-based electricity (0.26 $/kWh with 40 % subsidies on capital investment). In comparison with SHS, micro-grid-based

Table 4 Effect of subsidies on the electricity price

Subsidy as a percentage of capital cost (%)	0	25	40	50	60
LCOE ($/kWh)	0.35	0.29	0.26	0.24	0.22

electricity offers several benefits to the consumers. SHS can power only limited number of DC appliances including few CFL/LED bulbs, radio and a black and white television. If users need to power many appliances then a system having a large capacity PV system and a battery must be purchased and they are costly and are rarely affordable.

As mentioned in the introduction, off-grid hybrid system is one of the options among several for rural electrification. As for the case of grid expansion, it is not only expansion of the transmission and distribution lines, but also includes building new power plants since the present installed capacity is not sufficient for supplying new connections. Transmitting power from a central power station to a remote community involves losses along with the transmission losses. Further, another major problem experienced in rural areas with grid electricity is the low power quality (under voltage) due to losses along the long distribution lines. Hence, the relationship between the national grid expansion and off-grid electrification is quite unclear. In conclusion, we can say that off-grid hybrid systems are also as competitive as the grid expansion or SHS.

10   Grid Connection of the Hybrid System

The rural community in Sri Lanka selected for this study has not been electrified by the national grid yet. However, in the future the government may invest money for the extension of the national grid to this remote village as well. But, how long it will take to accomplish this task is uncertain. However, if the grid is extended to the selected rural village during the lifetime of the hybrid system, it will affect the cost returns of the project. Hence, developing off-grid project requires greater coordination between the private entrepreneur and the bureaucracy responsible for grid extension, which will help to design subsidy schemes and other policies to ensure sustainability of the system at the planning stage (Urpelainen 2014).

In this part of the study, we have investigated the impact of the grid expansion after several years on the economy of the project. For this analysis, the assumption is that the community will have access to the electricity from the national grid after 10 years. Under these assumptions, the community will buy electricity from the Independent Power Producer (IPP) during the first 10 years and then from the national grid. Hence, the IPP can sell the electricity generated by the hybrid power system to the community during the period when national grid is not available. Once the national grid is available, the hybrid system can be connected to the grid and all the energy generated by the hybrid system can be sold to the national grid. When the hybrid system is connected to the grid, the diesel generator and the battery bank will not be required any longer, thus they can be decommissioned. Then the hybrid system will consist only of renewable systems that is, 4 wind turbines with the total capacity of 40 kW and the PV system with the capacity of 30 kW.

The LCOE analysis of the project is made considering off-grid operation of the hybrid system for the first 10 years and grid connected operation for the next 10 years. All relevant costs including the initial capital investment, operating and maintenance cost, battery and generator replacement costs, salvage values and the earnings made by selling electricity to the grid after connection to the grid are taken to the LCOE analysis. Optimization in HOMER is done to find the amount of energy that the hybrid system can sell to the national grid. According to the HOMER calculations it is found that both wind turbines and PV system together can generate 157500 kWh of energy during a year. The purchasing cost of the energy is considered as 0.16 $/kWh, which is the current purchasing price of the energy generated from wind and solar by the CEB. The present value (C_{NPC}) of each cost (C) that will make n-years later is calculated by the following equation: (Kolhe et al. 2002)

$$C_{NPC} = C \left(\frac{1 + i'}{1 + d}\right)^n \tag{1}$$

It is assumed that the inflation rate i' is 5 % and the interest rate, d, is 7 %. It is found that the NPV of the lifetime cost of the project is $269000. In order to find the LCOE, the total net present cost of the project must be converted to series of equal annual cash flows, which is known as total annualized cost. The following equation is used to calculate the total annualized cost.

$$\text{Total annualized cost ($/year)} = \text{Total NPC} \times \text{CRF}, \tag{2}$$

where CRF is the capital recovery factor and it is given by the formula;

$$\text{Capital recovery factor} = \frac{i (1 + i)^N}{(1 + i)^N - 1}, \tag{3}$$

where i is the real interest rate and N is the number of expected years of recovering the total NPC of the project. The real interest rate is given by the following equation.

$$i = \frac{d - i'}{1 + i'}, \tag{4}$$

Once the hybrid system is connected to the grid, the generated electricity can only be sold to the grid at a known price defined the CEB. Hence, the present value of the earnings that will be made for the time period of grid connected operation can be calculated, since the annual energy production from the PV and wind systems is known. All other expenses (O&M cost, replacement costs) during 20 years of operation remain the same as the previous case. The income that will be made by selling the electricity to the grid has to be deducted from the investments in calculating the total NPC. This cost has to be recovered within the first 10 years of operation by selling generated electricity to the community because the earnings from selling electricity to the grid has already been taken into account. Hence, the number of expected years of recovering the total NPC should be taken as 10 years when calculating CRF, so the total annualized cost.

LCOE of the electricity generated by an off-grid hybrid energy project can be calculated from the following equation:

$$LCOE = \frac{\text{Total annualized cost} \quad (USD/yr)}{\text{Annual load served} \quad (kWh/yr)}, \qquad (5)$$

Based on the calculations, total annualized cost of the project is found as 30000 $/year and the LCOE of 0.3 $/kWh. This cost is less than the LCOE obtained for the case of off-grid operation of the hybrid system during the whole lifetime of 20 years. Therefore, it can be concluded that this system is economical to implement even though the grid will be extended to the Siyabalanduwa area in the future. That is the effect of connecting the hybrid system to the national grid after some time of off-grid operation affects positively on the cost returns. If the electricity is sold at a price, which is higher than the calculated LCOE before for the off-grid operation during the whole lifetime, then the project will make profits for both off-grid and grid connected operations (Kolhe et al. 2015).

11 Conclusion

This chapter presents a feasibility study of an off-grid hybrid renewable energy system for supplying electricity to a rural community in the Siyambalanduwa region in Sri Lanka. Siyambalanduwa area receives abundant solar irradiation and wind energy throughout the year. Therefore PV, wind with battery energy storage and diesel generator are selected as the candidate technologies and the optimum sizing of each component in the hybrid system is done based on the TLC analysis in HOMER. According to the simulation results, a hybrid system with 30 kW PV system, 40 kW wind system, 25 kW diesel generator 222 kWh battery bank is found as the optimum hybrid configuration and it can supply the electricity at a cost of 0.34 $/kWh. The annual capacity shortage of this system is only 0.1 %, resulting only 9 h of load shedding during a year due to the inability of supplying the load. The total NPC of this project is $553000 and it requires initial capital investment of $296000 to implement the project. The effect of the changes in the annual average wind speed and solar irradiation on the LCOE of this system is analysed. It is found that the increase in the LCOE for a decrease in the annual average wind speed to 4.5 ms^{-1} is only 0.1 $/kWh and the effect of change in the annual average solar irradiation within the range of 4–5 kWh/m^2/day is negligible. The analysis is done to find whether the system can supply the increasing demand. It is found that the optimized hybrid system can supply the increasing demand up to 320 kWh/day without any significant change in the LCOE, but the renewable fraction reduces as the load increases.

The total lifetime cost analysis of the project is made considering off-grid operation of the hybrid system for the first 10 years and grid connected operation for the next 10 years. Since the electricity generated by the hybrid system can be sold only to the community before connection to the grid, the total annualized cost is calculated considering 10 years. That is the first 10 years of operation of the hybrid system after commissioning the project. It is found that the annualized cost of the project is 30000 $/year and the LCOE is 0.3 $/kWh. This cost is less than the LCOE obtained for the case of off-grid operation of the hybrid system during the whole lifetime of 20 years. Therefore, it can be concluded that this system is economical to implement even though the grid will be extended to the Siyabalanduwa area in the future.

References

Annual performance report (2013). Ministry of Power and Energy, Sri Lanka.

Cho, J., Jeong, S., & Kim, Y. (2015). Commercial and research battery technologies for electrical energy storage applications. *Progress in Energy and Combustion Science, 48*, 84–101.

Farmann, A., Waag, W., Marongiu, A., & Sauer, D. U. (2015). Critical review of on-board capacity estimation techniques for lithium-ion batteries in electric and hybrid electric vehicles. *Journal of Power Sources, 281*, 114–130.

Grid interconnection mechanisms for off-grid electricity schemes in Sri Lanka (2013). Public Utilities Commission of Sri Lanka.

Households in occupied housing units in districts and Divisional Secretary's Divisions by principal type of lighting (2012). Department of Census and Statistics, Sri Lanka.

Kirubi, C., & Jacobson, A. (2009). *Community-based electric micro-grids can contribute to rural development: Evidence from Kenya. World Development, 37*(7), 1208–1221.

Kolhe, M., Kolhe, S., & Joshi, J. (2002). Economic viability of stand-alone solar photovoltaic system in comparison with diesel-powered system for India. *Energy Economics, 24*, 155–165.

Kolhe, M.L., Ranaweera, K.M.I.U., & Gunawardana, A.G.B.S. (2015). Techno-economic sizing of off-grid hybrid renewable energy system for rural electrification in Sri Lanka. *Sustainable Energy Technologies and Assessments, 11*, 53–64.

Microgrids- promotion of microgrids and renewable energy sources for electrification in developing countries (2008). Intelligent Energy-Europe.

Mohammed, A., Pasupuleti, J., Khatib, T., & Elmenreich, W. (2015). A review of process and operational system control of hybrid photovoltaic/diesel generator systems. *Renewable and Sustainable Energy Reviews, 44*, 436–446.

Nayar, C. V. (2010). Diesel generator systems. *Electrical India, 50*(6), 54–64.

Nehrir, M., Wang, C., Strunz, K., Aki, H., Ramakumar, R., Bing, J., et al. (2011). A review of hybrid renewable/alternative energy systems for electric power generation: Configurations, control, and applications. *IEEE Transactions on Sustainable Energy, 2*(4), 392–403.

Off-grid electrification using micro hydro power schemes- Sri Lankan experience, a survey and study on existing off-grid electrification schemes (2012). Public Utilities Commission of Sri Lanka.

Population of Sri Lanka by district (2012). Department of Census and Statistics, Sri Lanka.

Rolland, S., & Glania, G. (2011). *Rural Electrification with Renewable Energy: Technologies, quality standards and business model.* Belgium: Alliance for Rural Electrification.

Skyllas-Kazacos, M., & McCann, J. (2015) Vanadium redox flow batteries (VRBs) for medium- and large-scale energy storage. In *Advances in batteries for medium and large-scale energy storage* (pp. 329–386). Woodhead Publishing Series.

Urpelainen, J. (2014). Grid and off-grid electrification: An integrated model with applications to India. *Energy for Sustainable Development, 19*, 66–71.

Zaghib, K., Mauger, A. & Julien, C. (2015) Rechargeable lithium batteries for energy storage in smart grids. In *Rechargeable Lithium Batteries From Fundamentals to Applications* (pp. 319–351).

Printed in the United States
By Bookmasters